103
104
112
149
154-164

66
122

59 147
169

36 8
170 14
44

153

150
151

97
22-32
121
166

48

51

11
12

13

77

15
75
126
171

52
71
87
107

39
56
76
89
118

102
18

148
172

50
128
178

190 17

7

165

GRAND DESIGN

GRAND DESIGN

the earth from above

Georg Gerster, 1928-

PADDINGTON PRESS LTD

NEW YORK LONDON

My thanks go first, and with particular emphasis, to Swissair, and above all to its advertising manager Albert Diener and its art director Fritz Girardin. Commissions for aerial photographs for advertising and promotional use were only one form of encouragement I received from them; the second form derived from a shared conviction that the view from above could contribute a great deal to Man's understanding of himself. Without Swissair's stimulation this book would have got bogged down in its early stages.

My thanks go to the photographer Emil Schulthess. As the designer of a number of posters and calendars for Swissair using my aerial photographs, he unselfishly helped to focus the picture themes more sharply. Collaboration with him made my own eye more sensitive, and the book owes a lot to the standards he set.

My thanks go to the designer Hans Frei. Not shrinking from the unrewarding role of fill-in, he stepped in at a critical stage of the printing and made his immense knowledge available for the color corrections in the plate section.

My thanks go also to the National Geographic Society in Washington DC. The Society enabled me to undertake a reconnaissance flight along the Niger and permitted me to include some of the resulting photographs in this book.

My thanks go finally to all those who assisted me in the preparation, execution and evaluation of flights, namely Gianni S. Amado, Dakar; Ali Ghalib al-Ani, Baghdad; Emil Baettig, Mexico City; Behnam Abu 's-Suf, Baghdad; Rainer Michael Boehmer, Baghdad; Valerio Bondanini, Rome; Theodor Bregger, Milan; Helmut Brinker, Zurich; Peter Bürgi, New York; Donald G. Campbell, Sydney; René Caretti, Bucharest; Arnold Catalini, Baghdad;

W. H. Cook, Johannesburg; Reinhold Debrunner, São Paulo; Pio G. Eggstein, Johannesburg; Otto Eichenberger, Zurich; Peter Eigenmann, Hong Kong; Michael Evenari, Jerusalem; Catherine Flynn, Gove; Goro Kuramochi, Tokyo; Guy G. Guthridge, Washington DC; Derek T. Harris, Los Angeles; Carlos Häubi, Mexico City; Michael Heizer, New York; Sebastian Hirschbichler, Vienna; Thino W. Hoffmann, Colombo; James G. Holwerda, Singapore; Max Hunziker, Zurich; Thomas Immoos, Tokyo; Zacharias B. Kaelin, Sydney; Peter Kaplony, Zurich; Max A. Landolt, Washington DC; Fritz Ledermann, New York; Rudolf Müller, Tokyo; Walter Müller, Lagos; Peter R. Odens, El Centro, California; Erwin Walter Palm, Heidelberg; Roger Pasquier, Peking; Fritz Peyer, Hong Kong; Hans Röthlisberger, Uerikon/ Zurich; Hans Rüesch, Medan; Alfredo Schiesser, Philadelphia; Walter A. Schmid, Zurich; Hans-Jürgen Schmidt, Baghdad; Ernst Schumacher, Berne; Mario Selva, Athens; Adolf Senn, Rome; Robert Smithson, New York; C. G. G. J. van Steenis, Leyden; William Sturtevant, Wauneta, Nebraska; Elva Weakley, Calexico, California; August E. Weber, Hong Kong; Max Zimmerli, Kinshasa; Pierre Zoelly, Uerikon/Zurich.

Library of Congress Cataloging in Publication Data

Gerster, Georg, 1928—
    Grand Design

Translation of Der Mensch auf Seiner Erde
    Bibliographical: P.
    Includes index
    1. Aerial Photographic in Geography
    1. Title
G70.4.G4713   910'.22   76-3809
ISBN 0-8467-0140-5

© 1976 Georg Gerster
Translated by Renee Meddemmen
Edited by Stanley Mason
All photographs by Georg Gerster
Design: Heinz von Arx, Stäfa; Heinz Egli, Zürich
Separations & Plates: Photolitho AG, Gossau/ZH
Typesetting: Filmtype Services Ltd., Scarborough, England
Printing: Vontobel-Druck AG, Feldmeilen/ZH
Binding: Buchbinderei Burkhardt, Zürich
Printed in Switzerland
Published by Paddington Press Ltd., New York, London

Contents

Beautiful views

". . . in the plane give him many beautiful views and don't let the hyenas eat him . . ."
(from the nightly prayer of my then six-year-old daughter)

In the last few days of December, 1858, Gaspard Félix Tournachon, alias Nadar, embarked on the first successful photographic flight in a captive balloon; as a precautionary measure he had already applied, a few months before, for a patent on the utilization of aerial photography for the production of maps. On August 23, 1966, at 16.35 GMT, the ground control center in California radioed instructions to the photo-satellite Lunar Orbiter I revolving around the moon to glance back at its home planet. For the first time in history Man saw — even though only through technical servo-eyes — his own earth as a celestial body. After little more than a century, the domination of aerial photography had been superseded by the development of space photography. Prematurely, it seems to me: many possibilities of aerial photography from medium flying heights have yet to be explored.

Certainly, town and country planners, geologists and generals, tax assessors and realtors, cartographers, foresters, agricultural and civil engineers have long been making good use of aerial photography in their respective fields. For them, as for a number of other specialists, it is everyday fare and certainly no cause for emotion. If these specialists and technicians have no wish to get excited, that is their business. As I see it, however, an aerial photograph is a unique vehicle for wonder, vexation, joy, wrath — it never leaves one cold. To those sensitive to visual impressions it is a new training in observation, an unusual school of vision; to the concerned contemporary it is a mirror in which he can see himself as part of his terrestrial environment.

Today, nobody will naively ask a photographer to record the

world "as it is." In every photographic effort worth its salt, regardless of whether it is thought of as art, there is hidden an attempt to scratch off the patina of habit that covers everything like a sticky varnish, and to create the world anew. To do this, there is no need to play God: when the ego of the photographer comes to the fore and sets the scene, the result will be just as good as the ego; and that, most of the time, is provincial, artsy-craftsy and pretentious. The aerial view solves this problem at one stroke. The temptation to stage, to manipulate objects, is missing in the first place; moreover, the impossibility of meddling brings freedom: whatever it is that happens to be in the lens will not object. Above all, however, to the observer from above — I am thinking of the almost perpendicular angle, not the church-tower perspective — the world is re-created without tears, as fresh and unsullied as in Genesis.

Philip Gilbert Hamerton, an English art critic of the 1880's, defended the customary perspective against the angel's view. He mentally assessed the adequacy of the spatial and aerial picture by imagining the archangel Raphael descending to earth. The only landscapes he accepted in the artistic sense were the ones that could be seen from the ground; somewhat contemptuously, he left the aerial views to the geographers. This opposition can still be met with today, despite the fact that Raphael is now called Gagarin or Armstrong. "The bird's-eye view is nonhuman," a painter friend of mine once reproachfully told me; otherwise nature would have equipped Man with wings. But what about our dream of emulating birds in flight, and the gray matter that enables us to realize this dream? The visual transformation of things that flight bestows on us as a free bonus has always fascinated artists. The Italian Futurists anticipated landscapes of a kind never seen before from *aeropittura*, painting from the airplane — "as if they had just dropped from the sky." Henri Matisse in his old age apparently regretted not having been able to fly when he was young: the view from

10

above in his early years would have saved him the long detour back to his true self. Despite the elevated viewpoint, aerial photography—rather unexpectedly—does not create space, but surfaces: and Matisse wrestled with the organization of surfaces all his life. It is said that the master advised his younger colleagues to fly, for there was no better key to the optical code of the world. A new conception, then: flying not as the shortest route between two crowded airports, but as a shortcut to oneself.

It is hard to get one's fill from above of this second face of nature, the patterns and ornaments that Man, as he lives and labors, coaxes and bullies out of untouched nature. One look at the fields the farmer has laid like a patchwork quilt over valleys and plateaus offers a recipe for becoming an artist without really wanting to. The photographer, however, must be humble. What if he captures a gallery of Poliakoffs—or any other painter—on a single flight! The very fact that he sees them, when he sees them, is the merit of just those artists: they have helped to sensitize us to the forms and patterns that are now revealed in the hitherto hidden beauties of our world.

When my plans for an aerial view of the earth were beginning to crystallize, I still uninhibitedly indulged, as I flew, in a hedonism of the eyes. I was sometimes completely overpowered by beauty—it was always there, lying in wait for me, and I had nothing but an arsenal of cameras with which to confront it. The calligraphy of roads, the graphics of plantations, the unwitting art of desalination ponds and the mosaics of small cultivated fields still delight me and tempt me to board planes. But in addition to this beauty "out of the blue," and of equal importance, I am now aware of the information gained from the air. The aerial view by far exceeds the ground-level view in informational content; occasionally it even achieves something like the quadrature of the circle: the volume of information grows with abstraction. Admittedly, first doubts are stirred by the realization that even Man's worst offenses are aesthetically upgraded by sufficient distance. The automobile scrapyard in a natural setting is an eyesore on the ground, but even from kite-flying heights it is transformed into an attractive multicolored design. And as for the profuse, untidy settlement growth that eats into field, forest and meadow: at jet altitudes, if not lower, the eye begins to recognize a gratifying order in the chaos. Contemplation from a spacecraft redeems the earth from Man completely: to a lunar astronaut it appears as a habitable, though perhaps uninhabited, blue planet. This phenomenon of redemption through distance is the one drawback of an approach that otherwise has only advantages. Distance creates clarity and transforms the single image into a symbol: into an accusation here, a hymn of praise there, a manifesto everywhere. Coincidence turns to fact. On the ground we worry about an inventory of what is, but the lofty contemplation of the aerial photograph shows us also what might be — it is a stocktaking of our chances. Aerial photography x-rays the environments created by Man and reveals the intensity of the ecological give-and-take. It follows Man on his precarious way between foolishness and efficiency, conquest and coercion; manifests Man's conflict between the biblical order to subdue the earth and the necessity, only recently recognized, to submit himself to it. The currently popular condemnation of Man, which sees him as an incurable disease of his own planet, passes judgment without trial. I regard my aerial photographs as the interrogation of the accused; but if they plead at all, it is for one who has built up rather than against one who has destroyed.

So much for the "beautiful views." The hyenas my daughter wanted to deliver me from have accosted in many guises in the course of working on this book: the sloppy plane mechanic, the reckless pilot, the irresponsible aircraft owner — all of whom are exceptions in their trades. I think nothing of boarding a flying machine from the pioneering days of aviation, but I have learned to judge when a vintage plane is a year too old. I refused to be talked into flying in one that was pressed on me for a trip over the Atlas Mountains because of its untrustworthy age and looks. (And with some justification, as it later turned out: the owner had banked on a crash and hoped to collect the insurance money.)

My aerial photographs materialized at the price of a good deal of nervous stress. There was that near-collision with a high-speed helicopter over Stonehenge (Plate 168), which lies under a NATO corridor used for the testing of new equipment. . . . There was that flight above Kyushu in the forefront of a typhoon that had swept the sky clear of clouds. . . . Once we were ready to start at Mosul for the return flight to Baghdad, in the midst of Mig patrols taking off and landing. The control tower suddenly refused us permission to start, a jeep came racing toward us across the runway, and my heart sank at the prospect of the seizure of my films; but the driver smilingly presented me with a bunch of roses—"from the garden of the airbase commander, with his compliments. . . ." During a relaxed conversation on a flight back from the San Andreas Fault (Plate 2), the pilot began to tell me about the heroic days of aerial photography, about carrier pigeons and rockets used as camera carriers when he was a boy—shortly after the turn of the century! When I asked in alarm how old he was, he said he was seventy-four and didn't normally fly anymore, but had had to replace his son, who was busy elsewhere. I felt rather queer. I have a lot of respect for old age, but preferably on the ground. . . . On a flight over the crater of Erta Ale (Plate 23), dusk surprised us and the nearest air-field—without radio, radar, landing aids or lights—was 30 miles away and 9,800 feet higher, on the plateau. Things looked bad. The pilot, a chaplain and military aviator, had to rely less on his celestial connections than on his flying skill. During the landing approach I read the instruments for him, as he could no longer see them. Finally,

13

in the last glimmer of daylight, we touched down and, in complete darkness, taxied to a standstill on the uneven field. I had made it yet once more. . . . Then there was an emergency landing on the Panamericana, and routine landings on Brazilian airfields where cattle seem to have the right of way. I could go on, but it is not my intention to make the reader's flesh creep.

One's nerves actually suffer less in the air than on the ground, before the first takeoff. For days, sometimes for weeks, one waits for clear weather. Then the pilot of the small plane has to be cajoled into removing a window or a door — or, better still, both. He justifies his reluctance by talking of "safety," but he means "comfort." And almost everywhere one gets involved in battles with officials and authorities for whom aerial photography is synonymous with espionage. The regulations and prohibitions were formulated in the days when there were no satellites observing us uninvited from space. The problems raised by restricted areas cannot always be dealt with in the casual and friendly manner adopted by the authorities in one African state when they warned me against flying into forbidden zones. "As a specialist in aerial photography," the security chief reminded me in a fatherly tone, "you should know best what you are allowed to photograph." When eventually one can take off in a plane that has been stripped to the bone, there is often a security official on board. I have nothing against security officials in general, but a lot against those who object to every shot the photographer wants to take because they are at a loss to understand his passion for innocuous fields, colors and patterns. I have gradually learned, however, how many narrow circles it takes over an approved photographic target to make the watchdog feel sick, and I have discovered that a sea-green security official does not care much about security: he nods apathetically at each and every request. It is particularly satisfying to fly with pilots who see through the same eyes as the photographer, but I have learned to

14

fear those who are too easily infected by my enthusiasm: above the steep escarpment of Bandiagara in Mali we got into an ugly spin because the pilot insisted on taking photographs himself.

Wings, incidentally, are no guarantee that one will not need one's feet. One wintry day (−20°F) I took off from Columbus, Ohio, with a pilot who was unfamiliar with the area and we flew around in a vain search for Serpent Mound State Memorial, a prehistoric monument (Plate 184). Finally we landed in a field of stubble and I had to walk half a mile cross-country to reach the nearest farm. At first the farmer was rather nonplussed by the appearance of an alien, dressed in a light suit, who came out of the cold and asked after the Great Serpent. Only after making sure that I had really come from a plane and not from a flying saucer did he offer me coffee and information.

Legwork is also necessary when a place discovered from above subsequently has to be identified. During an exploratory flight along the Big Bend of the Niger, a village of rare beauty (Plate 35) suddenly appeared in my lens; months later I spent many days in the border area of Mali and Niger inquiring about its exact position and name. Labbézanga it is called, and the people clearly remembered the plane that had circled above their heads for hours (or so it had seemed to them). My interest in them and the fact that I had had a plane at my disposal at first aroused their suspicion; they thought I was a government agent sent to arrange for their threatened resettlement. When this misunderstanding had been cleared up, they joyfully hunted for their homes, their friends' houses and the village mosque on the photographs I had taken. Even though these Africans had never seen an aerial shot before, they had no trouble at all in reading my photographic plans of their village.

Most of the time I photograph from hired planes, preferably Cessna high-wing monoplanes. Small aircraft guarantee flying in the raw—raw as opposed to overdone, like a steak. For the commercial airlines nothing is too bad (films) and nothing too good (champagne) to make their passengers forget that they are in the air. Anyone who disturbs the routine of film and bottle by flattening his nose against the window hardly gains in popularity. But he should not let that worry him. Memorable pictures can sometimes be taken even through the double glazing of high-flying jets. As far as photography from passenger planes is concerned, I remember with melancholy the days before the epidemic of hijacking, when passengers could easily get admission to the flight deck. In the cockpit of the DC-3 there was even a sliding window that could be opened, and now and then an obliging captain would undertake an extra turn for a photographer. I would like to take this opportunity to apologize belatedly to the passengers of two commercial airliners. In the first instance a captain dropped 6,000 feet below the plane's cruising height to allow me to photograph the temple of Abu Simbel in Egypt; in the second, a flight commander kept his plane banking steeply over the rock churches of Lalibela in Ethiopia until my films were used up. In neither case did my fellow passengers know what was happening, and when I returned to my seat from the cockpit very few were able to lift their heads from their paper bags. Had they been up to it, they would gladly have flung me to the hyenas.

The earthly paradise, not excluding the Fall

Ours is a restless planet. When observed from above, the earth is in constant motion. Plates the size of continents wander along fault lines, molten rock oozes up from profound depths through the scars of the earth, water and weather furrow its face, celestial bodies bombard it with meteorites, hoofs abrade it, corals build whole mountains on it. And sometimes one species multiplies explosively at the expense of others, destroying inhabited and uninhabited environments and jeopardizing its own survival. I am talking, of course, of the lower animals. In the dynamics of nature Man is only one prime mover among others — and pictures of the earth from above give no support to his longing for a lost paradise. The Fall of Man is always included.

Two revolutions have put Man, once the crown of creation, into a more modest place. The one initiated by Copernicus displaced him from the center of the universe. He found himself the inhabitant of a fifth-rate planet circling around a fourth-rate, dwarfish sun, in its turn part of a mediocre galaxy. And the Darwinian revolution dampened his biological arrogance. He fell asleep created to the greater honor of God and woke up with an ape-like mammal in his pedigree. Both these definitions of Man's position are the achievements of a science that regards nature as a thing that can be manipulated and questioned; but both recommend, in the last analysis, humble integration in nature's order. Man's dilemma — to change nature or to adapt himself to it — is insoluble. Earth also belongs — a little bit, at least — to the elephants and the protozoa, to everything that flies or crawls. First of all, however, it is Man's earth, even though he may have no better claim than the higher specialization of his gray matter. The right of codetermination for wild animals? Partnership with all creatures? Democracy of living organisms? Fashionable models all, but foolish ones. Modern ecological thinking does not set out to question Man's despotism to this extent. But he must realize (and act accordingly) that in the community of living things he cannot tip the scales in his own favor without being guilty of deceit — and self-deception. The only thing we can hope for is that he will be an *enlightened* despot. . . .

1

Headland on the Sinai coast, in the Gulf of Aqaba, a continuation of the Red Sea, near Dahab. Virginal, eternal nature? Untouched by Man's hand, undoubtedly. But eternal—unmoving and unchangeable? The spring tide—rising here over six feet—has left azure pools and the gentle modeling of its flow in the sand. Along the shore corals build their reefs. And finally: the inlet of the sea against which this sand boomerang is flung belongs to a system of fault depressions extending over 4,300 miles from the coast of Mozambique to the Taurus Mountains. The rift valley of the Red Sea, and with it the Gulf of Aqaba, continues to widen measurably.

2

The San Andreas Fault in the Carrizo plain, California. It reaches the mainland at Fort Bragg, halfway between San Francisco and the border of California and Oregon, and leaves it in the Gulf of California, which is itself a widening of the fissure. Nowhere has erosion revealed the fault line more clearly than in the Carrizo plain. Horizontal dislocations along the fault have angled and displaced many dry stream beds. Along the line of the San Andreas Fault two huge plates of the earth's crust are in motion: the Pacific plate (on the right in the picture) with the coastal strip of California to the northwest, and the North American plate with the rest of the Golden State to the southeast. The annual displacement is only a matter of inches. Ten or even twenty million years may elapse before Los Angeles (on the Pacific plate) and San Francisco (on the North American plate) drift past each other. This mutual dislocation causes an accumulation of elastic tension which, as the citizens of San Francisco and Los Angeles know all too well, is eventually released in the form of earthquakes. A major earthquake (comparable to the one that destroyed San Francisco in 1906) is not only probable but 100 percent certain according to the prophecies of the experts, even though they are less than happy in their role of Cassandras. The only question is when, how and where; to several seismologists Los Angeles seems a more promising candidate for the next catastrophe than the city of the Golden Gate. For many young people who enjoy its sun and freedom California is consequently a kind of blessed nightmare—and the tectonic time bomb named for the holy Andreas is a guarantee that the nightmare will not last for ever.

3

The meteorite crater near Winslow, Arizona, U.S.A.; at the north end, a museum for visitors. It is the most famous of the "astro-blems"—the wounds in the earth's surface caused by the impact of bodies from outer space. The iron meteorite, estimated to have weighed 2 million tons, made a hole some 560 feet deep and 4,195 feet in diameter in the plateau of Colorado. Indians living in the vicinity have a legend according to which one of their gods descended from the sky with a thunderous din. It is actually unlikely, however, that humans witnessed what must have seemed like the end of the world: the meteorite fell some 22,000 years ago. The first white settlers took the hole for a volcanic crater, and only the pieces of meteoric iron lying around revealed their error. A Mr. Barringer of Philadelphia bought the crater outright in 1903 in the hope of exploiting a heaven-sent iron deposit. He drilled in the bottom of the crater for the giant meteorite—but in vain. At the estimated rate of fall the meteorite must have created a pressure wave that not only mechanically and chemically disintegrated the rock at the point of impact but also caused the meteorite itself to vaporize and burst into thousands of fragments.

4

River meander, seen from a plane between Fairbanks and Anchorage, Alaska. The first snow picks out the waterside meadows. The river reels from one side of the valley to the other, from slip-off to undercut slope and back again. The meanders are obviously shaped by the dynamics of flow, for no topographical necessity—such as debris deposits of tributary rivers—can be discerned. Here and there the river has cut through the neck of a loop. The terrain also discloses a few chapters of its past: dried-out oxbow lakes and silted-up tributaries.

5

Erosional filigree—a volcanic cone in Death Valley, California. The eruption of the volcano dates back a few centuries only. Since then, rains have carved grooves, furrows and gullies in the tufa. Gradually these expand into small ravines.

6

A miracle of weathering: Bryce Canyon in Utah, U.S.A. The term "canyon" is misleading, for this is the end-slope of a plateau of horizontally stratified limestone and sandstone with a few layers of shale and gravel conglomerates. For reasons not yet fully understood, there are zones in the sedimentary rock where joint planes run at right angles to the horizontal beds. In these clefts the forces of erosion attack: frost and plant roots burst the rock, water and chemicals contained in the air break it down, rain and melting snow wash away the detritus and continue the carving work on the rock towers and needles. Even their pigmentation is for the most part the product of erosion: during the chemical decomposition of the rock, red and yellow ferric oxide and lavender and purple manganese dioxide compounds are formed. This erosional wonder rightly enjoys absolute protection as a National Park. The man whose name it bears, however, does not seem to have made much of it. To the Mormon cattle breeder Ebenezer Bryce who settled here in 1875 the rocky labyrinth was "a hell of a place to lose a cow."

7

Centipede Reef, a part of the Great Barrier Reef, Australia, at low tide with little water coverage. Centipede Reef belongs to the outer reef of the coral bulwark 1,240 miles long lying in front of the northeast coast of the island continent. Among the coral reefs of the earth there is nothing that can match it; even in the earth's past, so geologists assure us, there has never been a comparable formation of corals. Lately Australia has been worried about its Reef. Natural forces over which man has no control are endangering the coral structures—earthquakes, cyclones and fresh water (when the Reef, laid dry at neap tide, is exposed to the rain). Man also increasingly interferes with coral growth by mistaken hydraulic engineering measures, water pollution and other ecological sins. Whether the massed appearance of the "crown of thorns," a starfish that feeds on the coral polyps, is also the consequence of an ecological miscalculation is at present disputed. The biologist Robert Endean holds that the extermination by Man of the

20

Triton's horn (*Charonia titonis*), a mollusk that is the principal natural enemy of the "crown of thorns," is responsible for the latter's sudden multiplication. The majority of scientists believe, however, that this circumstantial evidence is not enough to incriminate Man. Unlike Endean, they regard the mass attack of the "crown of thorns" as an episodic event which has also occurred in earlier eras.

8
Erosional pattern in the low plain between Euphrates and Tigris, Iraq. Rain and surplus irrigation water flow into a depression and take on the shape of a many-limbed dragon. The monster's color and oily glint are derived from minerals in the soil.

9
The U.S.S. "Glacier" breaking a channel in McMurdo Sound between Ross Island and the Antarctic mainland. The icebreaker does not ram the ice with its armor-plated bows but bears down on it until the ice bursts under its weight. A special device for heeling (pumping water from starboard to port containers and vice versa) even permits the ship to work its way forward on the ice if necessary, rather like a seal. The "Glacier" (8,600 tons, crew of 280, 21,000 HP) enables tankers and freighters to reach McMurdo, the biggest supply and research base in the Antarctic, one month earlier than if the natural bursting and drifting of the ice in the bay were awaited. Since the cessation of moon flights the Antarctic has again become the most unearthly ecosystem within reach of Man — and the most hostile; by far the coldest, driest, stormiest and highest. So far scientists have had the continent to themselves under a pact which in 1959 reserved the seventh continent for peaceful research and temporarily neutralized the territorial claims of the various nations. This arrangement keeps Man's interference to a minimum, and the Antarctic has remained a world almost as untouched as in prehistory. But how long can internationalism and virginity be guaranteed? Shortages of raw material and energy crises whet the nations' appetites for the mineral deposits that are known or believed to lie out there at the end of the world.

10
A piece of Ross shelf ice near its breaking point in McMurdo Sound, Antarctica. The ice is moving from the south toward the observer. Here, near the open sea, the glacier ice is already saturated with salt water. The characteristic surface texture in this transitional zone, however, has less to do with its salinity than with moraine coverage which, in the southern summer, causes different rates of melting. The Ross shelf ice, which is twice the size of Nevada, is an ice plate up to 320 feet thick that swims on the Ross Sea, rising and falling with the tide. Its inner velocity of flow is almost three tenths of a mile per year; at its edges it calves, giving birth to icebergs.

11/12
Flamingoes on Lake Hannington, Kenya. More than half of the world's flamingo population (estimated at six million) lives on and by the soda lakes in the East African rift valley. Two types of flamingoes wade, nest and breed in the inhospitable alkaline waters, which are avoided by other animals: the small flamingo (*Phoeniconaias minor*) and, in much more limited numbers, the large or common flamingo

21

(*Phoenicopterus ruber*). The beaks of both types are ingenious straining devices with which they are able to pick up the ample supplies of algae, tiny shrimps and other small crustacea without having to drink the poisonous alkaline solution. The plumage of the young birds is a dirty white which later turns pink as a result of their diet of salt algae containing canthaxanthine, a coloring matter chemically related to Vitamin A (see also Plates 93—96). Zoo visitors should know that flamingoes in captivity lose more and more of their pink color with every molting season unless their daily food contains paprika, carrots and/or a canthaxanthine preparation. The small flamingo drinks from the freshwater inlets of the soda pans, avoiding, if possible, the numerous hot springs on the shores of the lakes; these springs are responsible for the holes visible in the flamingo carpet. Untrained observers often assume that large congregations of a given species must be the result of living in a natural paradise. They could not be more mistaken. Although the flamingo has little to fear from other animals with the exception of the eagle and the marabou, it has the vicissitudes of the weather to contend with. Lakes dry out or are flooded at the wrong time, and either eventuality leads to abnormal movements in the annual migrations of the millions of birds (which are so far incompletely understood). Mass populations go hand in hand with mass extermination. The lakes become death traps when their salts crystallize out on the weak and spindly legs of the chicks.

13/14

Roads before their invention: paths of the gnus in the steppe of the Serengeti, Tanzania; and trails of horses, cattle, sheep and goats in the Mesopotamian lowlands, Iraq.

15

A "red tide" in the Bay of Ise on the Japanese main island of Honshu. Our environmental problems have led in many quarters to the assumption that, without Man, nature would always be right. Yet this red bloom on the water is caused by a breakdown of the ecological balance without any interference on the part of Man. Under changed local conditions — temperature, salinity and nitrogen supply among other factors — the dinoflagellates (armored unicellular flagellata in the ocean plankton) multiply with the momentum of an avalanche. The reddish streaks may be caused by marine phosphorescence produced when, at night, a ship or a fish induces the myriads of floating organisms (a type of dinoflagellate) to emit light. Or the "red tide" may be composed of dinoflagellates whose metabolic poisons are a serious threat to fish. Red tide catastrophes destroying huge numbers of fish are a periodic occurrence on the Spanish and Portuguese Atlantic coast, the coasts of South Africa, South America and Florida, the Pacific coasts of both Californias, Australia and Japan. Japanese scientists are foremost among the students of this water plague and hope eventually to eliminate it. The red tide also costs human lives: mussels which have fed on dinoflagellates and concentrated their poison may prove fatal to the incautious gourmet.

16

Ghost forest in Hawaii — not Man's doing. The eruption of Kilauea Iki in 1959 — with record lava fountains almost 2,000 feet high — deluged the forest on its southwestern flank with ash and slag. Many of the trees—mostly of the family of the *Metrosideros* (ironwood, related to the eucalyptus)—fell in the direction of the trade wind that carried this rain of death.

17

Ghost trees near Gove, Australia—Man's doing. The tree skeletons—eucalyptus and mangroves—stand in the red mud of an alumina works. The red mud is the residue left by the manufacture of alumina; the color is due to the ferric oxide content. The mud is dried in collecting tanks and can be put to further, though limited, use as an admixture in the manufacture of pig iron, in the chemical and building industries.

18

Tarry residues on a beach east of Lagos, Nigeria. These black temperature curves are inscribed in the sand by the breakers along the whole Gulf of Guinea, from the Ivory Coast to Gaboon. The oil and tar residues that make hundreds of miles of lovely sand beaches unfit for use are not by any means the result of occasional accidents, such as tanker collisions. They are caused by ships that, contrary to regulations or at least with complete disregard for the results, regularly dump oil residues in the sea. The international convention for the prevention of marine pollution by oil is lax: instead of issuing a general prohibition it merely lists forbidden zones. According to American calculations, 6 million tons of mineral oil flow into the oceans every year—most of it in the course of the ordinary operation of ships. If this continues, marine biologists warn us that the oceans will be devoid of life within a generation.

19

Water hyacinths in the Laguna de Bay, Philippines. The water hyacinth, *Eich-hornia crassipes*, is an escapee from South America. Flower lovers, charmed by the pale blue beauty of this water plant when in bloom, carried it with them into the tropical and subtropical waters of Central and North America, Asia, Australia and Africa, where, delivered from its natural enemies, it developed into a plague. Its capacity for propagation is unbelievable: it doesn't grow, it explodes! Theoretically, on the assumption of constant and ideal conditions, one single rosette with its stolons could mat into a carpet big enough to turn the whole water surface of the earth into a swamp in only two years. This waterweed swarms over navigation buoys, clogs ships' propellers, blocks river ports, chokes irrigation canals and immobilizes pumps. In recent times at least one positive quality has been discovered in the devilish weed: apparently it helps to purify poisonous waste water. . . . On the shores of the shallow Laguna de Bay, the largest freshwater lake on Luzon, main island of the Philippines, the water hyacinth is a common sight. How these round mats of *Eichhornia* are formed, however, remains unclear. Perhaps they can be explained in the same way as the balls of weed to be found on the flat coasts of the Mediterranean, which consist of neptune grass (*Posidonia*) rolled up by the breakers. The round green islands, which are not anchored, can seemingly change their position—perhaps after being pushed away by a fishing boat (two outrigger canoes and a fishing net are visible in the picture). Their roots then pick up the detritus from the bottom of the lake and carry it with them; hence the holes they leave behind when they move. (I owe this suggested explanation to C. G. G. J. van Steenis, Director of the Netherlands Rijksherbarium in Leyden.)

23

20

Weekend nomads in the "American Sahara," California. In the dune area between Yuma and Calipatria buggy enthusiasts, hopping over the crests of the dunes like beach fleas, engrave their designs in the sand. Camping trucks, the sand buggy in tow, gather together to form small settlements. One man's pleasure, however, is another man's vexation. Desert hikers, incensed by this noisy competition, are not the only ones who would like to prohibit the invasion. The boom in off-the-road vehicles, ORVs for short — 7 million forest, field and desert buggies, trailbikes and snowmobiles, with a doubling of their number anticipated within five years — is ecologically nefarious. Howling juggernauts purposelessly plow up beaches and forest floors, and desert soil, still scarred by the tank maneuvers staged by General Patton in World War II as practice for the North Africa campaign, is finally delivered over to total loss of vegetation and to erosion. How can this environmental destruction be stopped? By prohibitive and restrictive laws, of course — provided they can be carried through against the lobby of pleasure seekers, vehicle manufacturers and misguided tourist interests.

21

Sea lions and sea elephants on "New Year" Island, off the California coast. The U.S. Coast Guard operated a lighthouse on Año Nuevo Island from 1873 to 1948, when the construction of a beacon on the mainland, six miles further up the coast, made the use of the island unnecessary. When the lighthouse keeper and his family vacated the station, thousands of seals moved in. The permanent inhabitants of the island now comprise over fifty species of birds as well as Steller's and Californian sea lions and sea elephants. In the fall the population temporarily increases tenfold: up to 15,000 Californian sea lions break their migratory voyage on the island. The pinnipeds have taken over Man's estate, inhabiting the ruins and wreckage he has left behind. For once his fellow creatures have had the last word. . . .

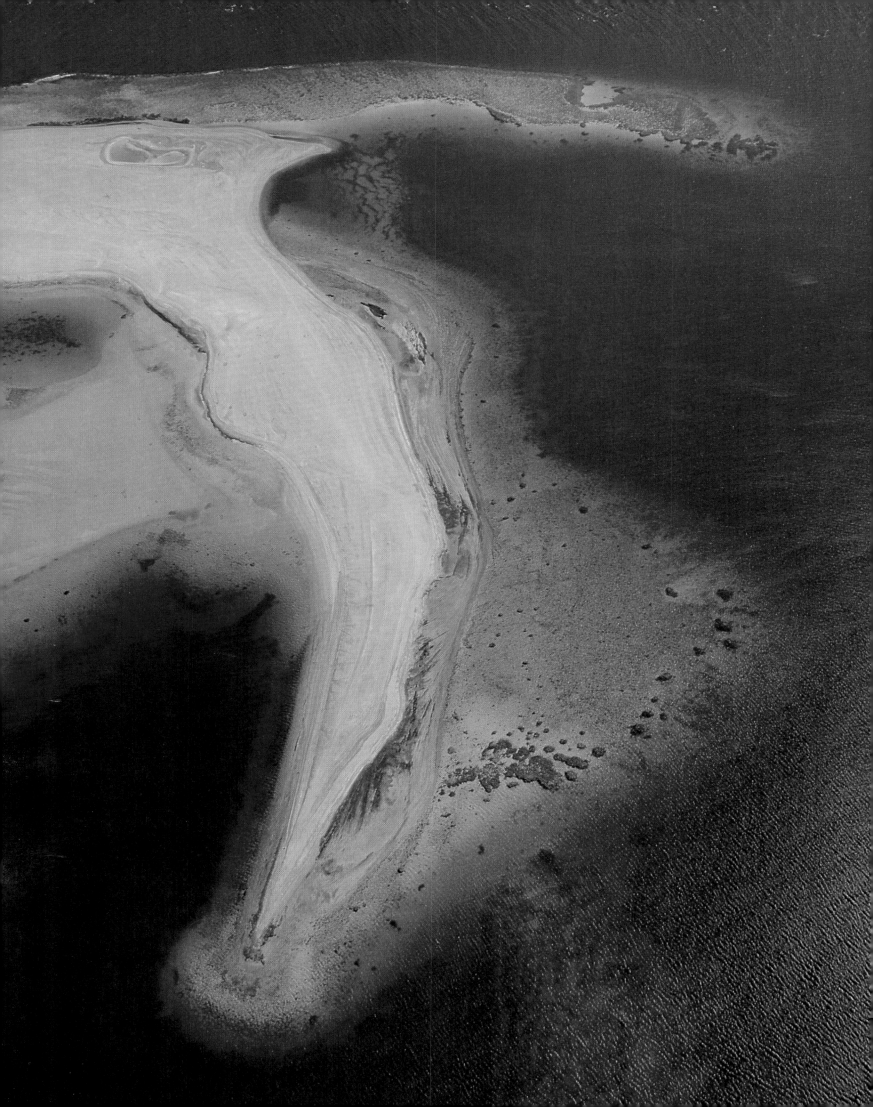

2   3
4▶

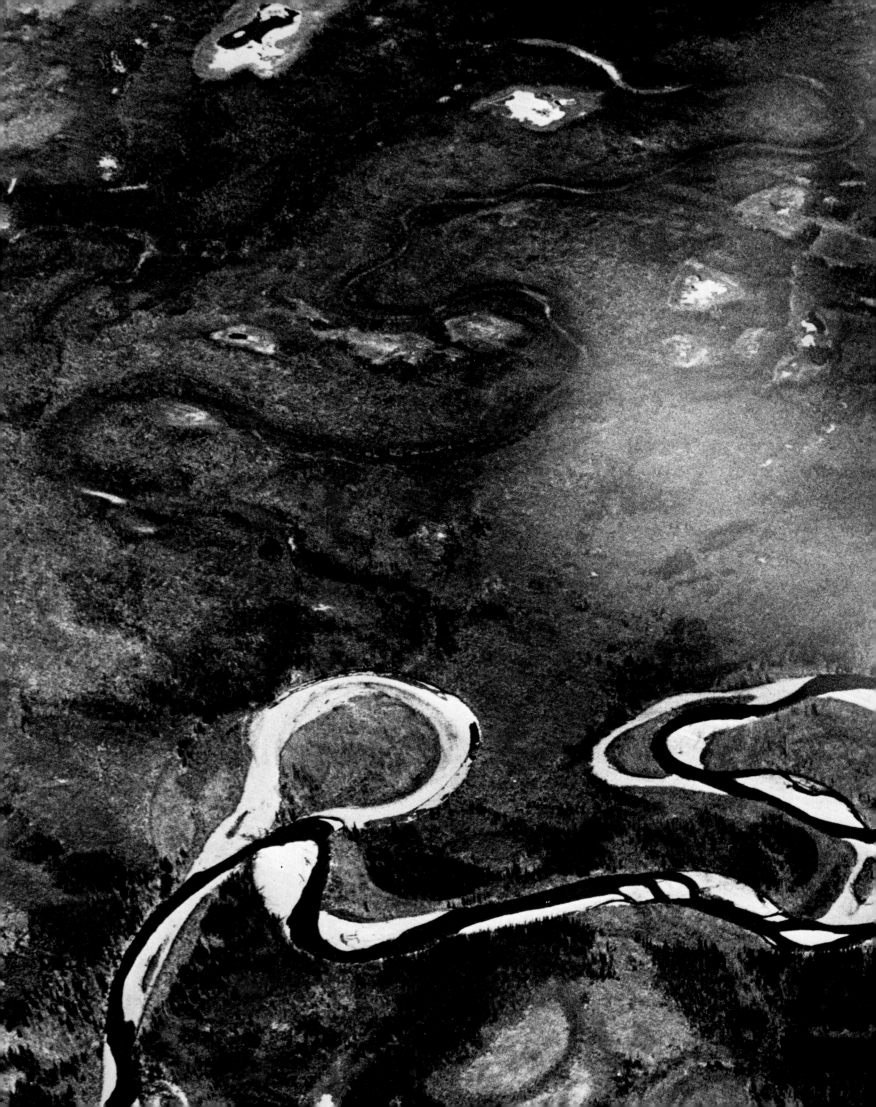

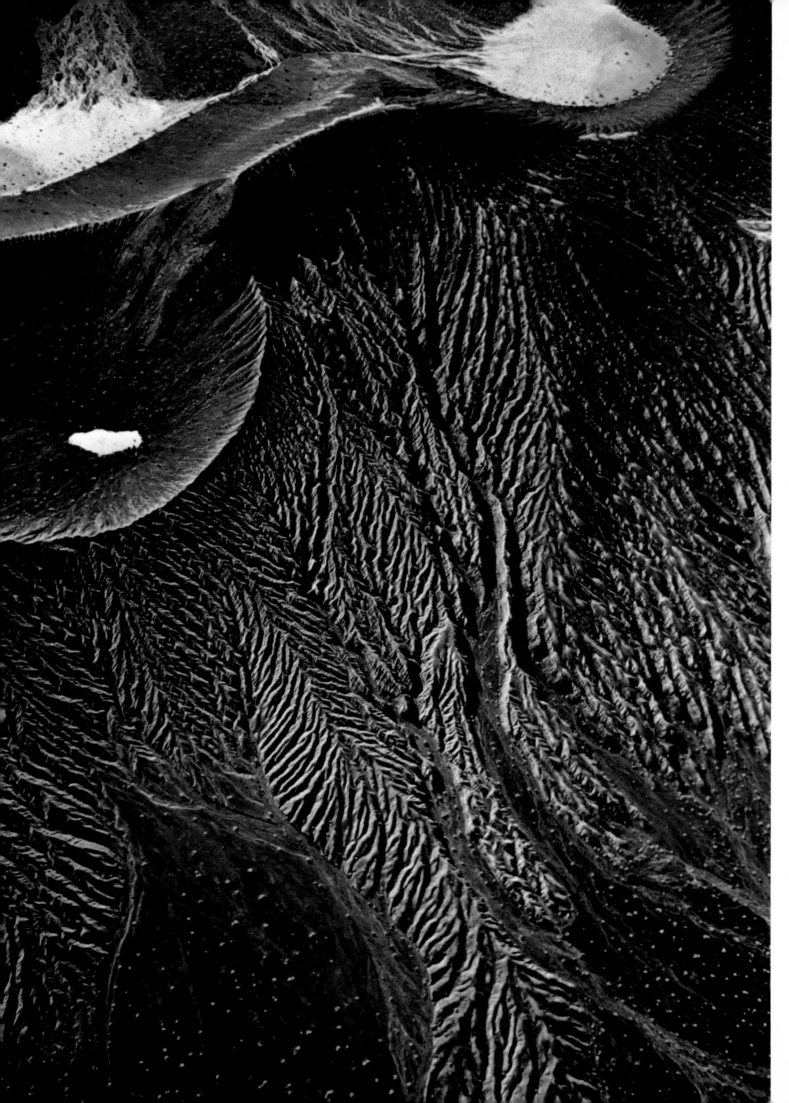

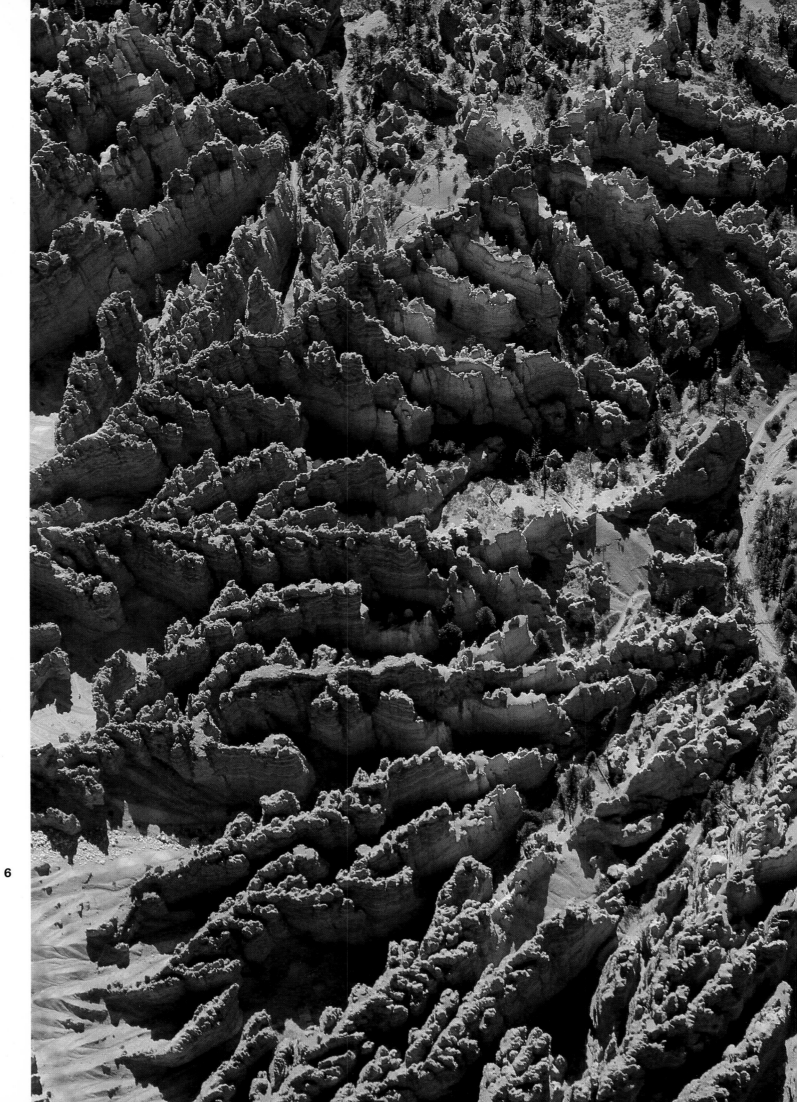

6

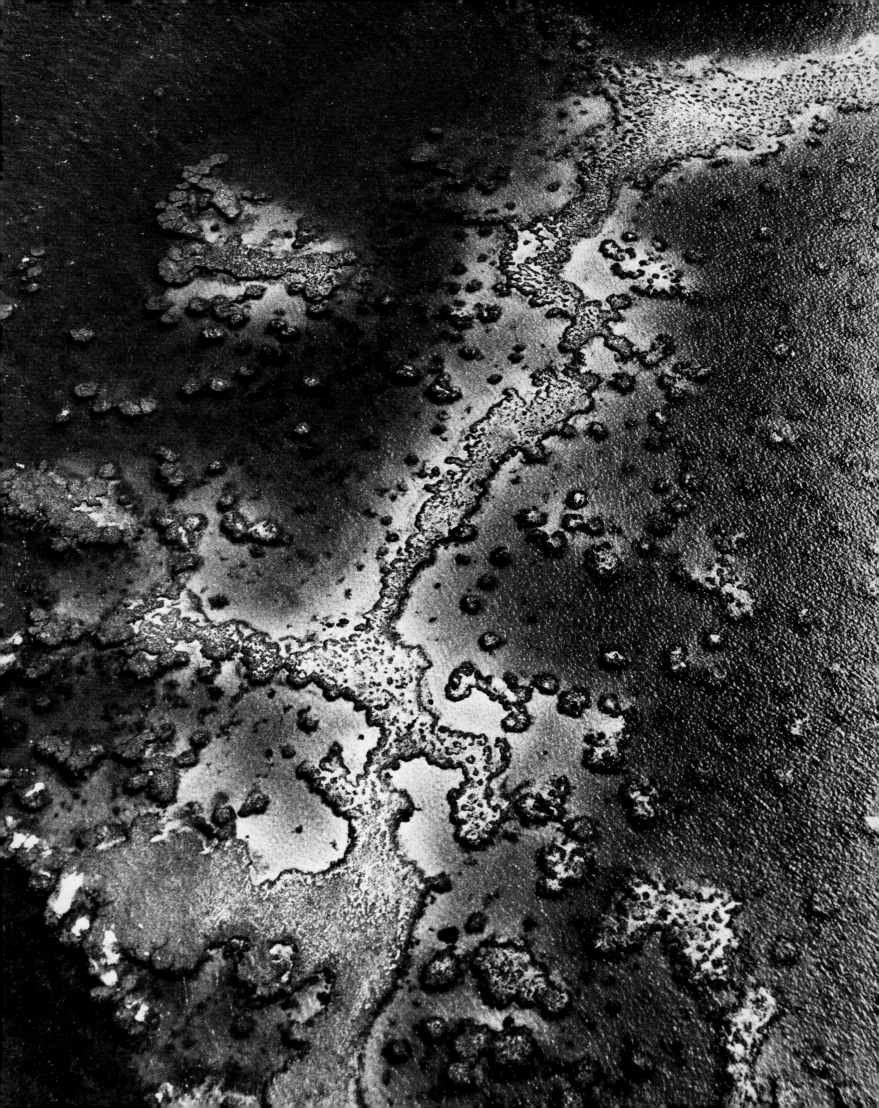

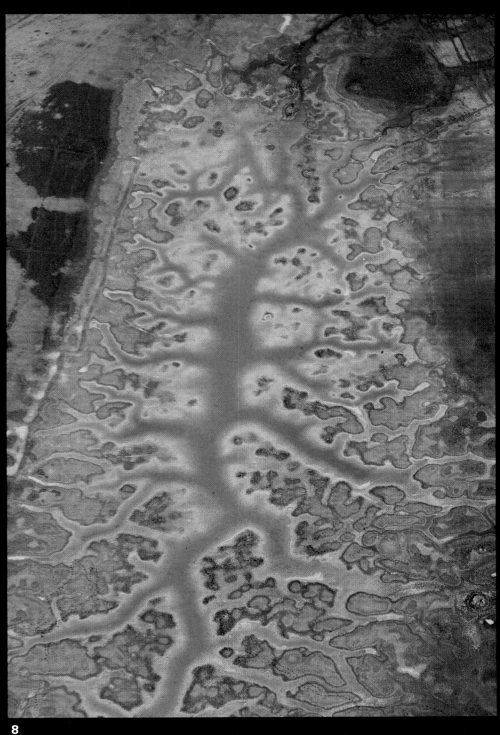

7   8

9

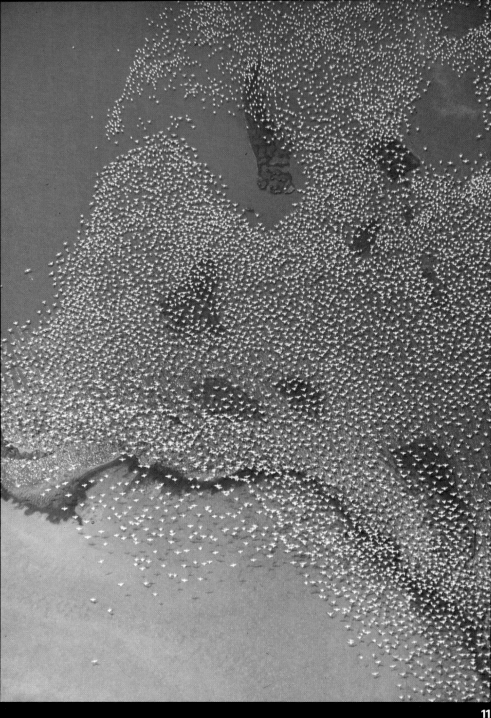

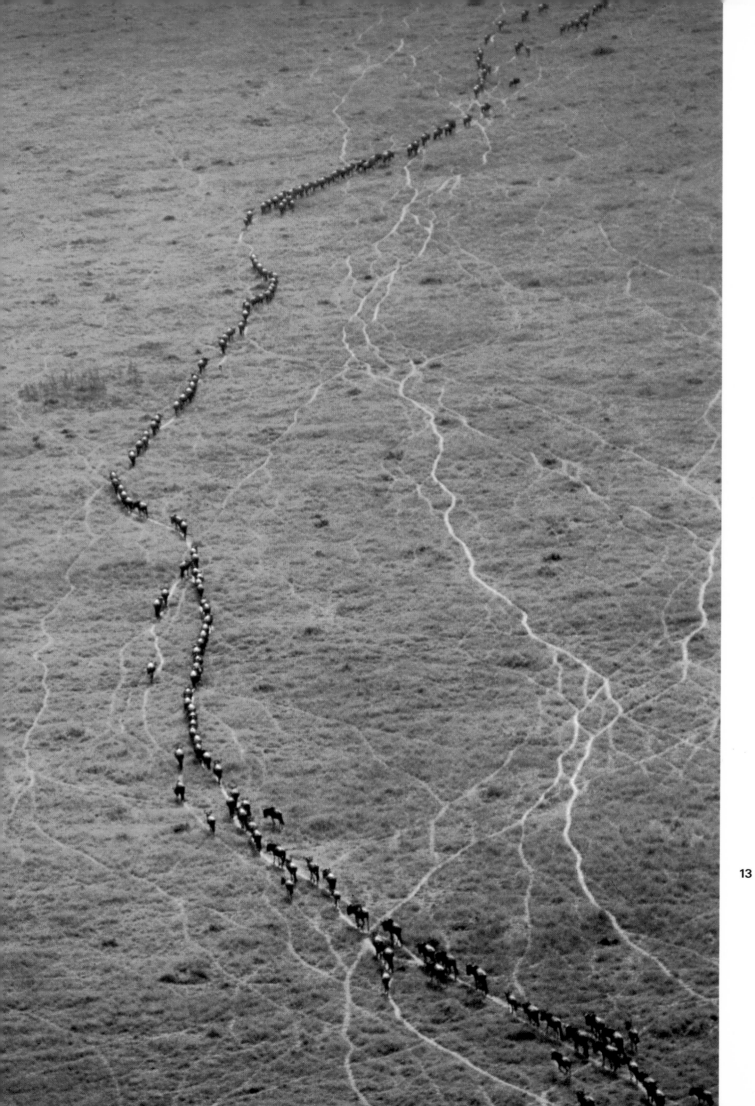

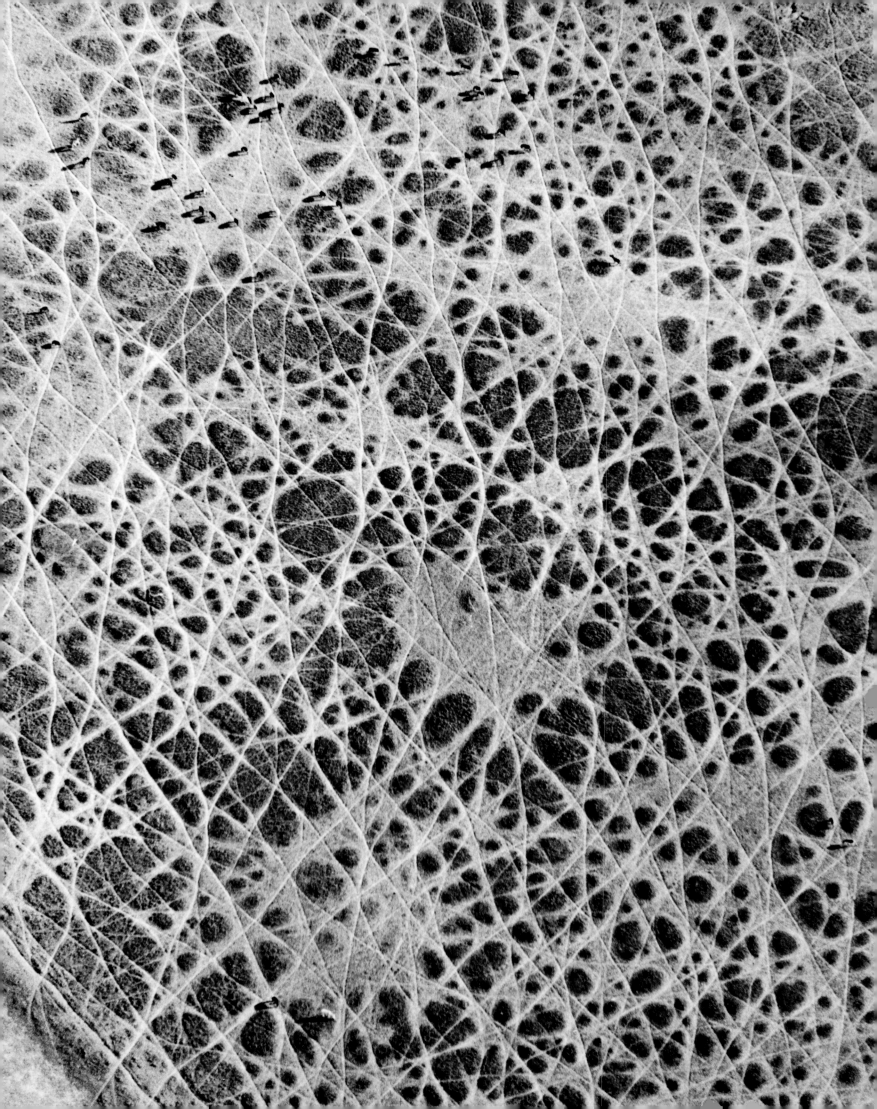

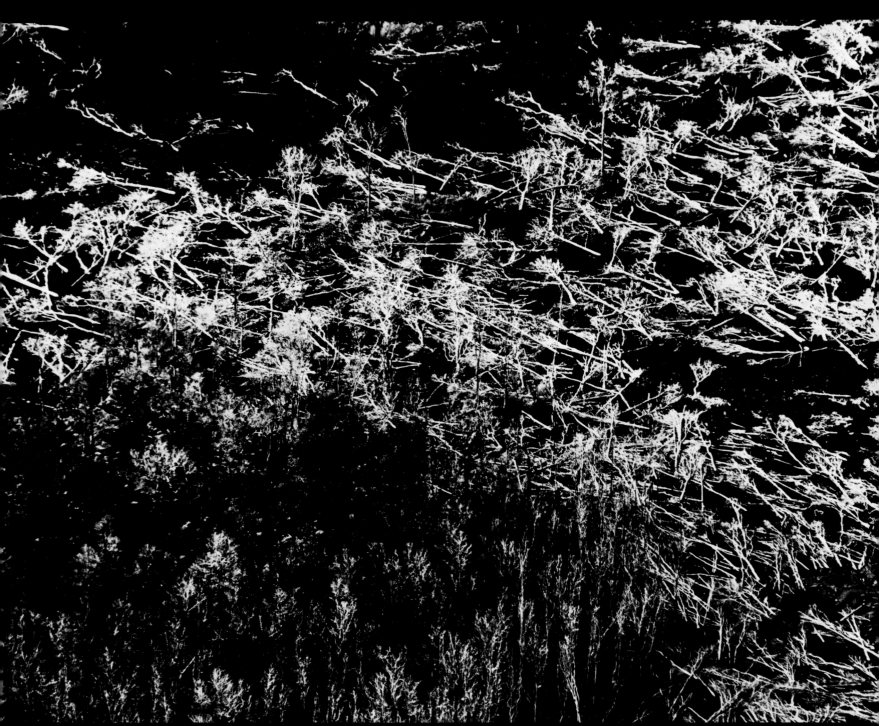

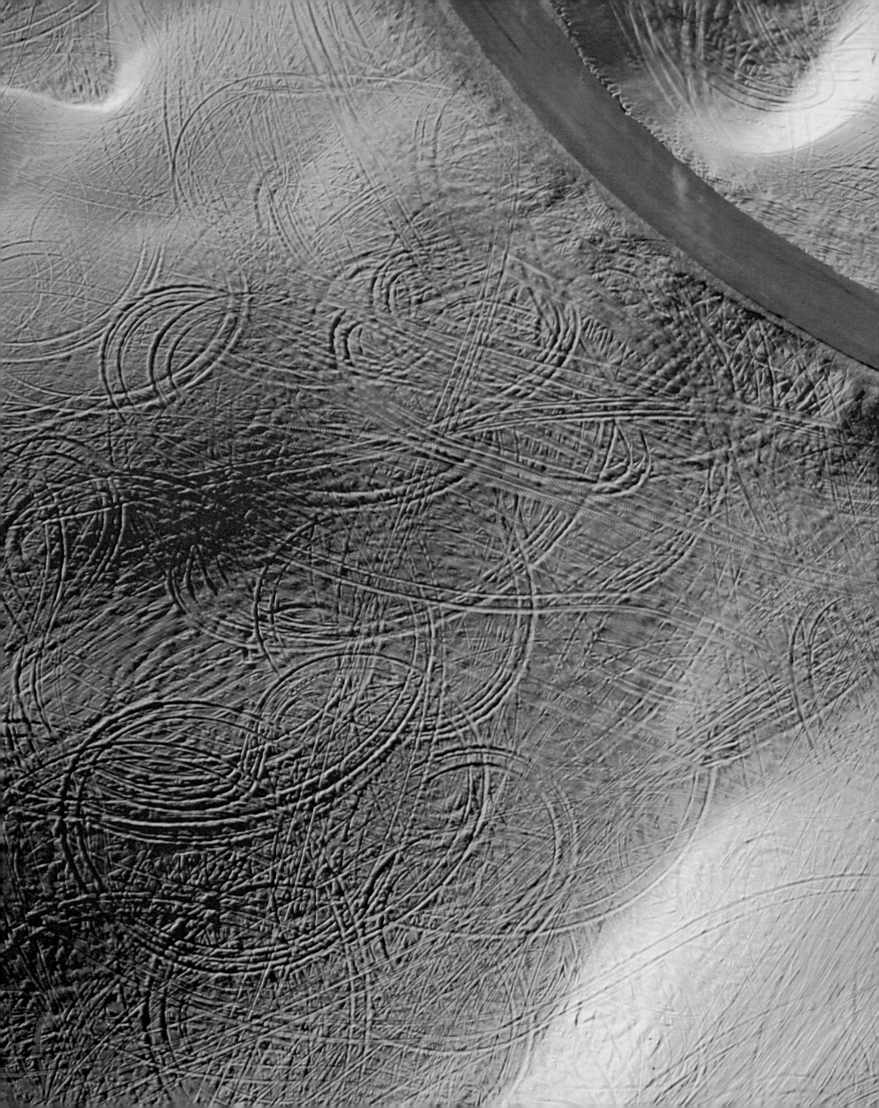

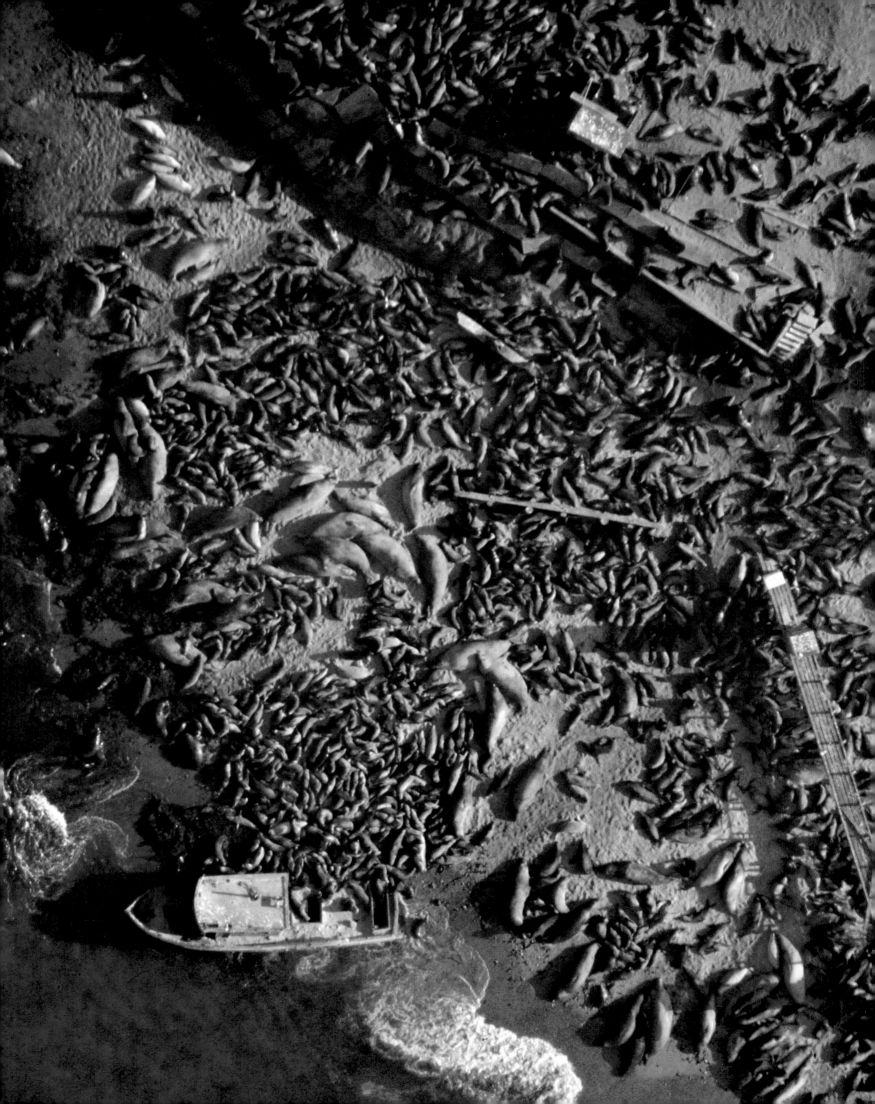

Flight over the Afar

Landwards the Afar triangle is bounded by the precipitous escarpment of the Ethiopian plateau and the terraced slope of the Somali tableland. Seawards it is flanked by the Red Sea and the Gulf of Aden. This desert, three times the size of Switzerland, belongs mostly to Ethiopia but overlaps in its southeastern corner with the territory of French Somaliland and the Somali Republic. The few thousand nomads traversing it—members of the Danakil tribes—can hardly be considered a population. The catastrophic droughts of the last few years have reduced their numbers drastically and driven them to the edge of their territory, where the climate is a little more favorable. *Afar*, meaning "the free," is the name the Danakil give themselves and has recently led to a new designation for their triangle of land.

For the traveler with no concern for the geosciences the Afar desert is an inferno, particularly in the north where it sinks below sea level. This is the planet's thermal pole: in the summer the mercury column creeps towards 140°F in the shade. Bleaching skeletons mark the way through this desert without hope or horizon; salt grinds underfoot and salty sweat stings the eyes. Then come oceans of dunes, lava slag and cinder fields, geysers that periodically spew steam and water, fumaroles and solfataras, and air reeking of sulfur. A vision of hell—and gaudy as sin: in Dalol yellow salt flowers grow out of a green brine, the chlorates of potassium and iron blossom into unearthly gardens. All in all, it is the apex of inhospitality.

But this is not the only reason that scientific research has been held up for so long in this part of the world. Danakil custom requires that young males should decorate their brides with the genitals of slain enemies. The fear of castration deterred even those who might have been willing to risk their lives, and geologists and geographers preferred to avoid the Afar desert. For the last eight years, however,

21

49

this African valley of death has attracted swarms of now unquailing scientists, for the Afar triangle might prove to be the Rosetta Stone of a new geological discipline, plate tectonics. (As it happens, palaeoanthropologists have also recently begun to see their Mecca in the Afar desert. The discovery of "Lucy," a unique prehistoric skeleton—3 million years old and 40 percent intact—makes the Afar a "cradle of Man.")

The science of plate tectonics rehabilitates Alfred Wegener and his hypothesis of the continental drift. Yet, according to this new Disney-like conception of the constant changes taking place on the face of the earth, the continents no longer plow through the earth's crust but travel as stowaways on huge plates constantly forming and disintegrating. Plate tectonics has identified six major plates of planetary proportions and, depending on the theories of individual scientists, hundreds of minor ones. Processes taking place in the earth's interior keep these plates in constant motion. Through the V-shaped median valleys at the summits of submarine ridges magma oozes forth, forming new oceanic crusts and pushing the plates apart. Where such plates collide (in the zone of the deep sea trenches), the heavier plate is forced down into the molten rock of the earth's crust.

An example of these tectonic uplift processes is to be found in the Afar desert, where just such an oceanic mountain, whose magma springs add new material to the plates, has gone aground. Here it is possible to study dryshod phenomena that otherwise can only be investigated by diving and drilling from submarine research vessels. Iceland is a similar lucky strike for geoscientists. The Afar triangle, however, is of particular interest to the scientist because several seam zones of the earth's crust meet here at a single point.

The frequency of earthquakes and intense volcanic activity are evidence of the tectonic lability of this area. Beneath the desert there is enough heat to supply the whole of Africa with electricity, and the red glimmer of the volcano Erta Ale lights up the night sky of the northern part of the desert. The "Smoking Mountain" (as the Danakil call it) and the Nyiragongo (Kirunga) in Zaire are at present the only volcanoes on earth with a permanent lava lake. Erta Ale is the only instance of a continuously erupting volcano in the median valley of a dried-out oceanic ridge. An aerial photograph I took of Erta Ale in 1965 (the first known aerial view of this volcano) later helped volcanologists to assess the extent of its movements: in the space of a few years the lava in the two craters rose 460 feet and eventually overflowed.

In the Afar desert three fault systems meet in the form of a star: the trough of the Red Sea, the continuation of the Carlsberg Ridge in the Gulf of Aden, and the East African rift valley. A large number of French, Italian and German geologists as well as some British, American and Ethiopian scientists are interested in this "rift star." According to the Belgian volcanologist Haroun Tazieff, spokesman of a French-Italian group, the trough of the Red Sea and the continuation of the Carlsberg Ridge in the Gulf of Aden are two sections of the same rift valley, which bends round in the Afar. His theory is based on arguments derived from surface geology. His colleagues, who plead for two independent systems, substantiate their disagreement with geophysical measurements, mainly seismological and obtained from explosions. It is certain that the Danakil depression in the north of the Afar triangle is an arm of the Red Sea which has been cut off by uplift processes. Here the salt lies to a depth of thousands of feet, and the potash deposits in this evaporating pan are estimated at millions of tons. On the lava floor of this enclosed sea basin — dry now for tens of thousands of years — corals can still be picked. Many peculiarities of this triangular rift valley system are yet to be explained. Wegener had already noted that the southwestern corner of Arabia and the African coast lying opposite fit each other like two pieces of a jigsaw puzzle if the Afar triangle is disregarded, but even today it is difficult to explain it away on scientific grounds. Parts of the desert

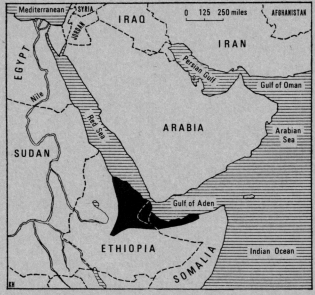

lie well above sea level; and some horsts of continental crust are troublesome stumbling blocks. Just as the bursting apart of the earth's crust between Africa and Arabia once created the Red Sea, in a few million years the Afar triangle will be flooded again (Tazieff calls it the "Erythraic Sea" in anticipation). But opinions on the present stage of its "oceanization" differed greatly in the spring of 1974 at an international Afar Symposium in Bad Berg-zabern in the Palatinate. Tazieff and his group already seemed to see the magma rising in all the clefts and fissures, preparing the way with its basalt for a future ocean. The German geophysicists, however, were more reserved; their measurements suggest a thin but still intact continental crust.

There are still many such unsolved riddles. From a flight across the Afar depression in 1971 I brought back a photograph of a swirling labyrinth of salt blocks. Later on, again from the air, I tracked down two further examples of such labyrinths, albeit badly disfigured by erosion. Geologists who had worked for years in the vicinity of this salt lake had, to my astonishment and their own, never come across these mysterious formations. For the moment they disagree as to their meaning—in fact, they are completely at a loss.

52

22

The two craters in the caldera of the Erta Ale volcano in the Afar depression. The diameter of the bigger crater is 980 feet, that of the smaller one 230 feet. Erta Ale is, technically speaking, a basaltic shield volcano in the process of formation. The picture was taken in 1965, when the lava lake was at low tide. In the bigger crater the level of the lava is approximately 460 feet below the crater rim; steam and smoke prove that there is also a lava lake in the smaller crater.

23

Erta Ale in spring, 1974: the big crater has filled and narrowed into a magma lake 165 feet in diameter. The molten lava forms a thin skin where it is exposed to the air. Where this skin breaks, hell blazes and bubbles.

24

Cone of an extinct volcano in the Erta Ale range. It was formed when the Afar was still an arm of the Red Sea; by the time it was left high and dry, it had already become dormant. Fish live in the crater lake, and the fringe of vegetation indicates that not even the blistering summer sun can dry it up completely. A natural cistern collecting the rainfall? Hardly, since rain here falls scantily and sporadically. There must be some connection with groundwater reserves; only an underground water supply can make up the losses due to evaporation over long periods of time. Prospectors baptized the structure "The Colosseum." Later it received the more official name of "Mount Asmara."

25

A curious ring consisting of multicolored rock salt and fragments of basalt. The geologists take it for a salt plug encased in rock, which surged up from the depths and broke through the surface. Its present structure is the result of erosion by rainfall and floods. The Afar people attribute special healing powers to the salt of this plug.

26

A stillborn volcano, so to speak: a magma intrusion from the depths of the earth which did not reach the surface but threw up a whirling labyrinth of salt blocks in the overlying strata. The striking rectangle at the center could be the remainder of a sealing block. Erosion transformed the labyrinth to its present state by washing away the more easily soluble salts and thus filling the spaces between the blocks up to the level of the surrounding, hexagonally textured salt plain. The "vortex" has a diameter of 490 to 655 feet. My attempt at interpretation here is based on information obtained from the geologist Derek T. Harris, who was a member of the geological team that investigated the potash deposits of the Afar depression, estimated at hundreds of millions of tons. Harris had never actually seen the vortical labyrinth (which was only six miles away from the prospectors' base on the hill of Dalol) but he associated my aerial picture with several enormous magnetic anomalies of similar shape appearing on the magnetic map of the area. These anomalies point to intrusion bodies with a high iron content, many of which, however, have produced no sign on the desert surface. Harris planned a hoax with my picture: he wanted to publish the photograph of the labyrinth in a reputable scientific journal as an archaeological site, as vestiges left by Man. His premature death prevented him from carrying out the plan. (His hoax might easily have backfired: James G. Holwerda, another geologist with years of Danakil experience and a colleague of Harris', seriously believes this vortex to be the remains of a camp of salt gathering Danakil nomads. Holwerda, too, judged from the aerial picture. Only an actual inspection of the site can clarify the matter.)

27
A bizarre world of pure salt: the hill of Dalol in the northern part of the Afar. It rises 150 feet above the salt plain lying nearly 400 feet below sea level. Magma surging from underground thrust up the hill, which is dotted with geysers; occasional rainfall has melted the salt blocks to towers and needles. Caps consisting of gypsum, which is harder to dissolve, protect the tops of the salt excrescences.

28
A tumult of colors in the salt desert: the rounded summit of the hill of Dalol. Hot springs (at about 212°F) shed their salts; the colors—including yellow, which might suggest sulfur efflorescence—are due to the oxidation of small iron admixtures. Each spring flows for approximately one month; once it has dried up, the deposits also gradually fade away.

29/30
Springs at the edge of the Afar depression, where the groundwater horizon cuts the slope. The groundwater is fed by the rainfall in the Ethiopian highlands as well as by underground water emerging along fault lines. The latter can be recognized by a temperature above the ambient level.

31
The monsoon rains in the Ethiopian highlands and the very scanty local precipitation bring some water to the Afar depression. The traces of the run-off water are engraved in the black "desert varnish," a crust of manganese. The diagonal running across the picture, which at first looks like a scratch, is in reality a caravan trail—a trace of Man among the veining of natural forces.

32
A salt caravan on its way back to the plateau. Exploitation of the salt deposits by hand has been for centuries the only source of income for the Danakil tribes in the north of the Afar triangle. The lowest point of the salt pan never dries out completely, and the monsoon rains over the highlands nourish it anew each year. Twenty thousand pack animals—camels, mules and donkeys—journey to and fro on the ancient salt road. The annual output is about 17,000 tons of table salt.

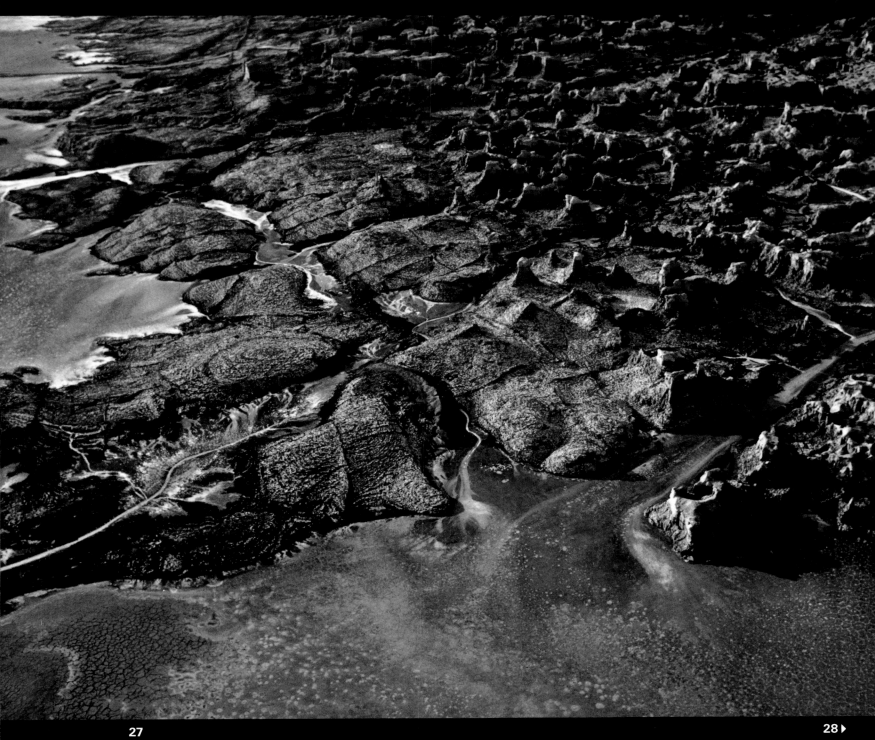

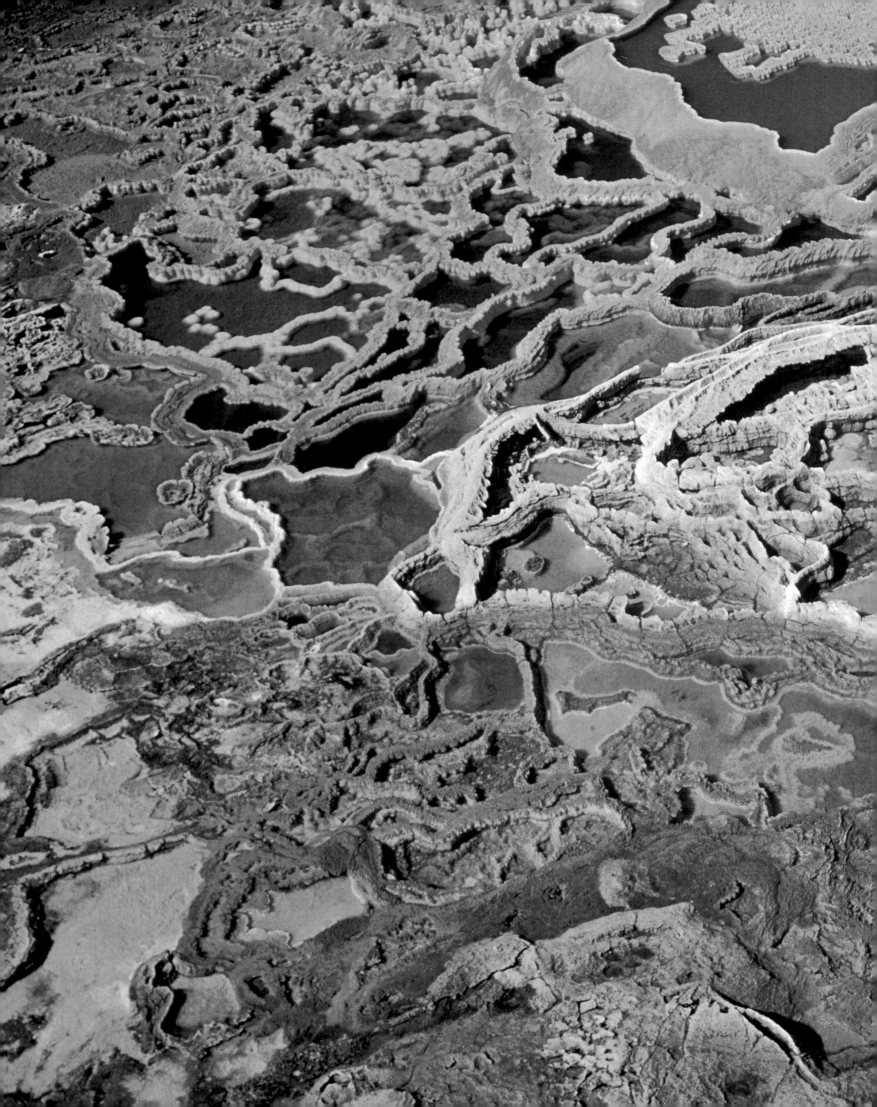

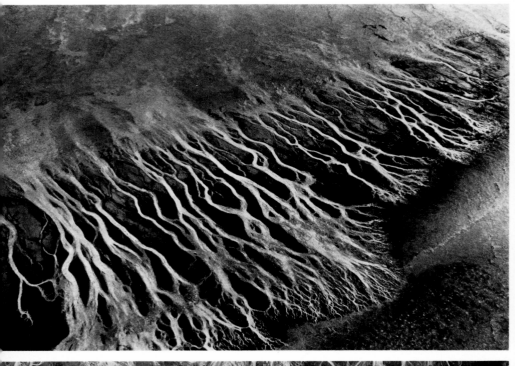

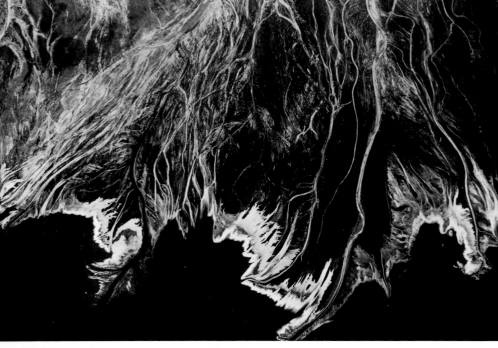

29-31

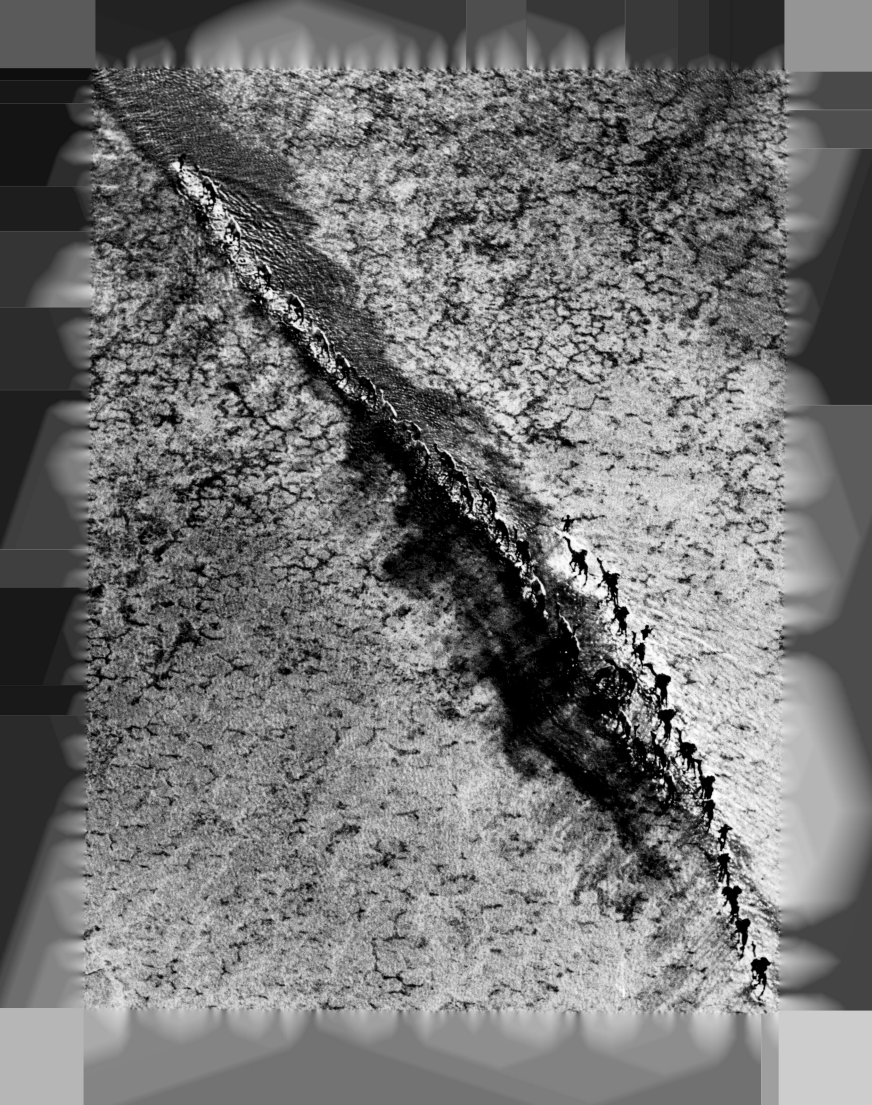

A roof over one's head

Aerial photographs showing vegetation on special color film (with colors that are not true to life) enable the biologist to distinguish between sick and healthy plants. This is now common knowledge even in the school classroom. The application of the same principle of diagnosis-from-the-air to settlements is—much to Man's disadvantage—not yet part of the curriculum. The view from above expedites analysis and affords insights difficult to obtain from the ground. For example, a half-hour flight over the Swiss plateau is enough to demonstrate how the cancerous proliferation of today's urban settlements destroys the landscape. In days gone by children with whooping cough were cured by sending them up in planes. Nowadays similar flights should be regularly prescribed as a shock therapy for town planners and citizens alike. *L'avion accuse,*" Le Corbusier noted in 1935. "The airplane is an indictment. It brings a charge against the city. It also brings a charge against those who control the city." The flier's eagle eye sees the cities as cruel, inhuman creations. It reveals "that men have built cities for men, not in order to afford them pleasure, contentment and happiness, but to make money!" Le Corbusier's vision of "Profitopolis" was directed against the big cities of the nineteenth century, as a contrast to which he cited the new architecture of the machine age.

Forty years on, alas, the airplane still indicts. The accused is no longer just the nineteenth century, but also the architecture and urbanism of the machine age: the worship of function and purpose, the reduction of the need for a home to a mere right to a watertight shelter. . . . The plane accuses. But it can also defend—and acquit. Although Le Corbusier had visited the valley of the M'zab in the Algerian Sahara on foot, it was only after flying over the area that he was overwhelmed by the exemplary nature of Mozabite towns—"a wonderful lesson" he called it. Since then our varied examples

of unintentional urbanism, of planning without planners, of anony-
mous, spontaneous architecture have been increased a thousand-
fold. We could never have found them, except from the air.

33

A masterpiece of medieval urban architecture: the old part of the town of Berne, Switzerland. It represents in unequaled purity and maturity the urban pattern of an axially structured town with a street market. In Berne the wider main street that served as a marketplace—Kram—and Gerechtigkeitsgasse, one of the most astonishing and congenial street systems anywhere—is flanked by smaller streets running parallel to it. The eaves of the houses are aligned and their gables form long rows. In front of the ground floors of the houses are the famous arcades: citizens were permitted to build these arcades before their houses and to extend the house façades over them; the ground under the arcades, however, remained public property. Berne is one of a whole family of towns founded by the dynasty of the Zähringers on both sides of the Rhine. The urbanist Paul Hofer, the leading authority on Berne, praises it as the "most compact and consistent creation of town planners, in the twelfth century at any rate."

34

The oasis town of El Oued, capital of the Souf in the Algerian Sahara. A showpiece of "architecture without an architect"— the dream of every town planner. . . . Vaults and domes covering all living and service rooms shed the wind-blown sand, which settles as a sound-damping layer on the streets. The house with courtyard is the unity of the town, which grows on the building-block system without any waste of land. El Oued, which has 50,000 inhabitants, is the most important market and administrative center of the Souf. A hidden stream flowing under the sand determines the form taken by date palm cultivation in the area (see Plates 109 and 110). But the architectural originality of the townscape is also connected with this stream. Water rising by capillarity evaporates before it reaches the surface, and the salts dissolved in it are thereby precipitated. Thus in one zone are formed banks of almost pure gypsum, which, when roasted, can be used as mortar and for making bricks in molds. In another zone the gypsum becomes encrusted with quartz sand to form sand roses and later *lus*, the local name for a hard, durable brick which sets extremely well with the gypsum mortar. This explains the brick construction of the Souf houses, which is very unusual for the Sahara. The clay generally used for building is not available in the Souf, and timber is if anything even rarer than elsewhere in the desert— hence the domes on the houses, the simplest form of roofing where timber is too costly. The workers erect these semicylinders and hemispheres by hand, disdaining any frames or shuttering. Although in this part of the world the dome is the privilege of religious buildings and of palaces, in the Souf the poorest and least holy of men lives in this respect like a marabout or pasha. Only the very well-to-do can afford a two-story house with a common or garden flat roof!

35

The village of Labbézanga on an island in the Niger, Mali. The storehouses for millet and rice (Plate 108) wind through the village landscape like strings of beads. These amphora-shaped storage buildings, some of them as high as the huts, are filled and emptied through an opening at the top; stone slabs and fragments protruding from the body of the storehouses make them easier to climb. The villagers, who belong to the Songhai tribe, still live for the most part in the traditional round adobe huts with hemispherical straw roofs. (The other traditional house form of the Songhai is shown in Plate 53.) Even in Labbézanga, however, the typical square house of Islamic-Arabic architecture is gaining in popularity: to live in one boosts the owner's social standing. A border village between Mali and the Republic of Niger, Labbézanga should have been moved to the river bank several years ago—a measure dictated by the government in Bamako for no more legitimate reason than official convenience. The villagers, looking at the ruins of a fortress of the once powerful Songhai empire on the island and remembering the defiant spirit of their ancestors, refused to obey the order from Bamako, and the government did not act upon its threat to evict them by force. Allah be praised for both! I know no more beautiful village in the whole of black Africa.

36

A farm equipped for self-defense to the south of Baghdad, Iraq. These small fortresses, which are occasionally intended as places of refuge for whole settlements and mostly belong to the sheikh, are known locally as *qalaa*. The rectangular walls have bastions at two corners so as to open the flank of attackers to those defending the farm.

37

Burg Kreuzenstein, a fortress near Korneuburg, Austria. A Burg Kreuzenstein already existed in the twelfth century but was destroyed by the Swedes during the Thirty Years' War. About the end of the last century it was rebuilt as a model of a Romanesque-Gothic fortress. Medieval building material gathered from the whole of Europe was used for the purpose.

38

A walled farmstead of the Kirdi near Rumsiki, in the Mandara Mountains, Cameroon. Huts and barns are huddled together for protection inside a stone wall. Such farmstead forts—known as *sarés*—lie like hedgehogs in a landscape of rural toil and travail (Plate 105), synonyms of the fear that drove the Kirdi up into the comparatively safe mountains. As animists (Kirdi means "unbelievers"), they owe both their fear and their name to their Mussulman neighbors who wrested the fertile plains from them.

39

A Chinese farmstead in the hinterland of the Kowloon peninsula, Crown Colony of Hong Kong. In these large farmsteads whole clans live together in very confined conditions. The ground plan of the farm is classical; comparable architecture can be found in pictures dating back to the Han dynasty (206 BC to 220 AD). Today the defenses serve only to ward off demons and evil spirits.

40

A village on the lake of Kainji, a storage lake of the Niger in Nigeria. The villagers are busy pounding millet and cooking breakfast as a new day begins. The picture hints at the delicate balance between "public" and "private" property, one of the basic features of African living. Stake and mat fences offer no protection against armed attack; theoretically rather than materially, they keep out "nature," safeguarding the clan's territory against wild beasts and the evil eye of strangers. A loud call for "authenticity" is now heard in Africa. No return to genuine African traditions can be complete without the acceptance of native forms of dwelling and settlement; in spite of the not very durable building materials, they greatly facilitate a deeper understanding of human communities.

41

Saint-Malo in Brittany, France. The Bretons, a saying has it, are born with seawater around their hearts. If that is true anywhere, it is in Saint-Malo. The Malouins are famed as explorers and seafarers: it was one of them, Jacques Cartier, who first reached the St. Lawrence and gave the territories he took possession of for France the name of Canada. And it was the Malouins who, as corsairs — officially sanctioned freebooters with letters of marque from the king — harried the English, Spanish and Dutch on the high seas. This territory smaller than the Tuileries produced many great men; its most famous citizen, Chateaubriand, seemed more satisfied than astonished at this fact. The Malouins are a proud and self-assured race, and their island town is an architectural gem which they have worked on since 1144, when it became an episcopal seat. Recent reconstruction — true to history both in spirit and appearance — has cleared away the extensive destruction wrought by World War II in this hotly contested town.

42

A suburb of Seattle, Washington — a paradigm of American suburbia. The well-tended residential estate, a monument to the good neighborhood (everybody in the same income bracket), and a breeding-ground for status-symbol culture and the seeds of racism. Among all the arguments advanced for the suburb, the automobile is the most dangerous: four-wheeled mobility intensifies the centrifugal forces that are bringing the hearts of American cities depopulation and death.

43

Downtown Los Angeles. The car-conscious Californian megalopolis has become the embodiment of the anti-town: a swarm of parking lots in search of a city. In the background the section of the Santa Monica Freeway that carries the highest traffic density (175,000 cars per day on eight lanes). More urban freeways will be found in Plate 70.

44

Islands and barrel-type houses of the Madan in the swamp wilderness on the lower reaches of the Euphrates and Tigris, Iraq. Where the two rivers meet lies the waste of reeds and water known as Hor, for centuries a place of refuge for the flotsam and jetsam of the tribes, with many pre-Arabian usages and customs. The only means of transport of the swamp-dwellers are the *meshhufs*, beaked boats with a high waterline, which they punt with great skill through the reed thickets. They live on their buffalo herds and on the sale of rush mats. Their semicylindrical dwellings are made wholly of reeds, and the islands themselves are also built on a foundation of reeds and other canes. In the course of the years, fed with the dung of the water buffaloes and the waste from the huts, these islands grow till they reach a height of several feet above the winter low-water level. Hydraulic engineering projects on the upper courses of the rivers, particularly the evening-out of high and low water in the interests of irrigation, today represent a threat to the thousand-year-old way of life of the Madan.

45

Dwellings on piles in Lake Nokoué, Dahomey. Lake Nokoué—also known as Lake Ganvié, after the biggest village—is the lagoon behind the sand bar on which Cotonou, the country's capital, stands. Fifty thousand people live as fishermen on the lake. They use net and harpoon, but also practice a kind of fish farming by putting out twigs for the fish to spawn in. Fishing is the preserve of the men, who sell their catch to their wives, who in turn sell the fish (mostly smoked) on the market. The lake dwellings are hardly older than the present century. After a channel had been cut through the sand bar in 1895, the freshwater lagoon became salty, and the increasing salinity made irrigation of the land impossible. The farmers were forced to become fishermen and moved out into the lake.

46

Kayar, Senegal. For seven months of each year life here is centered upon boats and fish; for the remainder of the year the village at the water's edge is deserted. In December it grows into the biggest fishing settlement on the West African Atlantic coast, only to die out again in June, at the end of the fishing season. Beyond the barrier of foaming breakers, rich catches attract the fishermen. The art of riding the waves in pirogues seems almost to be inborn in these West African fishermen. Yet the women and children still await the return of their menfolk with anxiety when, coming home each evening, fathers, husbands and sons must negotiate the dangerous wall of surf.

47/48

Zulu settlement in the "Valley of the Thousand Hills," near Durban, South Africa;

and Nilotic farmsteads along the Sobat, a tributary of the Blue Nile, Sudan. The round hut with a conical roof is the typical farmer's house in black Africa. The herdsmen usually lives in a hive-like or hemispherical hut without cylindrical walls. Originally the Nilotic tribes were solely cattle breeders, and the Zulus were strongly influenced by them. But the form of dwelling chosen by both of these tribes reflects the growing importance of agriculture in their lives.

49

The Edmundshof farm on Parndorf heath, Austria. The manor belongs to the Cistercian abbey of Heiligenkreuz. Parndorf heath, a sheet of gravel in Austria's Burgenland, is wind-swept and dry. On the occasion of a renovation seventy years ago, the farm was surrounded by a double circle of trees as a windbreak and was replanned with dwellings, stables and outhouses so far apart that a fire would not burn down the whole farm.

50

Oil palm plantation in North Sumatra, Indonesia. The settlement of the plantation workers forms a bright square. It is the germen of a company town and belongs to the prescribed settlement forms in which self-interest and goodwill—here that of the plantation company—are intermingled. In the foreground and middle distance young plantations are visible. For technical harvesting reasons the palms, which begin to bear in their third year, are only allowed to grow to a height of fifty feet, which takes twenty-five years. One twenty-fifth of the plantation area, therefore, has to be renewed each year. In North Sumatra, at the present time, about 180,000 hectares (nearly 450,000 acres) of land is under oil palms, with 143 trees per hectare. One palm yields around 225 pounds of fruits per year. Every harvest worker is expected to cut a ton of them every day in his block — about fifty bunches of fruit, each with hundreds of plum-sized drupes rich in oil.

51

Parts of the Konso settlement of Buso in the province of Gamugofa, Ethiopia. The Konso are among the most interesting of vestigial tribal societies in southern Ethiopia. They live in three dozen permanent settlements, the compactness of which is a feast for the eye of the town planner but a nightmare to epidemiologists. The population density of Konso settlements, which are towns rather than villages, is attained nowhere else in Ethiopia (and probably nowhere in the rest of Africa). The modular unit is the single farm with huts for living, sleeping and cooking, a storehouse and a stable. The single farms form quarters on the basis of family and clan, community organizations and ritual groups. Each quarter has at least one *mora*: a festival ground with a men's house in which married and single males sleep so as to save energy for war and hunting through sexual abstinence.

52

Part of the city of Nagasaki in Kyushu, Japan. In the period during which Japan cut herself off from the outside world (1639 to 1859) Nagasaki played an important role as the only port in Japan where Dutch and Chinese vessels were allowed to call. Thus, it sometimes acted as an exchange center for Western goods and Western science. One of the most nefarious products of the latter, baptized

"Fat Boy," was later responsible for the fact that today Nagasaki is all new: on August 9, 1945, at two minutes past eleven, an atomic flash obliterated 150,000 people and their city. Today heavy industry is concentrated in Nagasaki, which exports ships (see Plate 87). Since World War II a weakness for roofs with colored tiles, particularly blue, has spread over the whole of Japan.

53

A Songhai village near Timbuktu, Mali. The more than 500,000 Songhai live on the Great Bend of the Niger. They fish, cultivate a narrow strip of land along its banks, and harvest wild rice (Plate 108). Where mobility is required by their way of life or the terrain, they use exclusively a hut made of matting which resembles the shell of a tortoise. In this they follow the shoals of fish or escape high tides by moving to elevated land. But even when they settle down permanently they are loth to renounce the habits of their seminomadic past. Those who can afford to keep up with progress may acquire an adobe house with a flat roof, or at least a rectangular walled courtyard. Within it, however, the tortoise-shell huts still appear, either as accommodation for guests or for the hot summer nights. The building of these matting houses is women's work, and they are therefore the wife's property. A man who leaves his wife also loses his home. Plate 35 shows another type of dwelling used by the Songhai.

54

A camping town near Copenhagen, Denmark: the city dweller's frustrated dream of unspoiled nature.

55

Mobile homes in Long Beach, California. At bottom left, part of the concrete canal grandiloquently labeled "Los Angeles River." The trickle in it only lives up to its name in flash floods. For the privilege of parking his motorized dwelling the owner pays a monthly rent. Most mobile home estates, however, are not satisfied to be mere parking lots for portable houses: they embellish themselves with communal facilities such as clubhouses, saunas and swimming pools, recreational centers, golf links and gymnasiums. This extra glamor glues their tenants to the location, once chosen. One estate may offer personal wine cellars, another will hold a Rolls Royce at its clients' disposal for transfers to and from the local airport. Their calculations seem to be working out. Every year Americans buy over half a million mobile homes, and are proud to uphold the covered-wagon tradition. But once they have found a good site, they stay there— in fact, they move house less frequently than apartment dwellers.

56

Junks in the port of Aberdeen, Hong Kong. A hundred thousand people live on 21,000 houseboats and lighters stationed in the natural harbors and typhoon shelters of the Crown Colony. They are born on their boat and they die on it—very rarely do they set foot on *terra firma*. Aberdeen has the lion's share of Hong Kong's floating population. For the great feasts of the Chinese calendar almost 13,000 junks will meet in its port, and whole clans are anchored in an intimate circle, bow to bow. The inhabitants of the junks make their living by fishing, by small transport services within the harbor area or along the coast, and as traveling tradesmen. Because of its high, flat stern and unfavorable ballast distribution the junk is not very well equipped for the open sea. Junk enthusiasts will not admit this, but even the Hong Kong Tourist Office— certainly no enemy of things picturesque— warns visitors of the disadvantages of these boats out of official concern for their safety.

57

Brasilia—the reverse of the medal. One million people have been attracted by the nucleus of the new city, but such multitudes had never been reckoned with. A ring of squatter settlements (here known as *invasões*—"invasions") grew up around the city, which had been conceived as an architectural and urbanist showpiece. Glass, steel and concrete on the one side; iron sheeting, roofing felt, old crates and rags on the other. The battle order in which the "invasion" advances makes mockery of town planning by aping some of its principles. Despair packed in corrugated iron? Oscar Niemeyer, Brasilia's star architect, resigned in the face of these hovels, shacks and sheds: "In the end, Brasilia was a town like the others, a town of rich and poor, unjust and discriminating." Contemporary urban sociologists, however, make distinctions. They agree that not all "invasions" (and not all *bidonvilles*, shanty towns, *barriadas* and *favelas*) are slums. Slums are dead ends of misery, traps of desperation—including the comparatively well-heeled ghettos of North America and the noble slums of Europe. By contrast, squatter settlements, despite their unattractive building materials, may also be places of hope, scenes of a counter-culture, with an encouraging potential for change and a strong upward impetus.

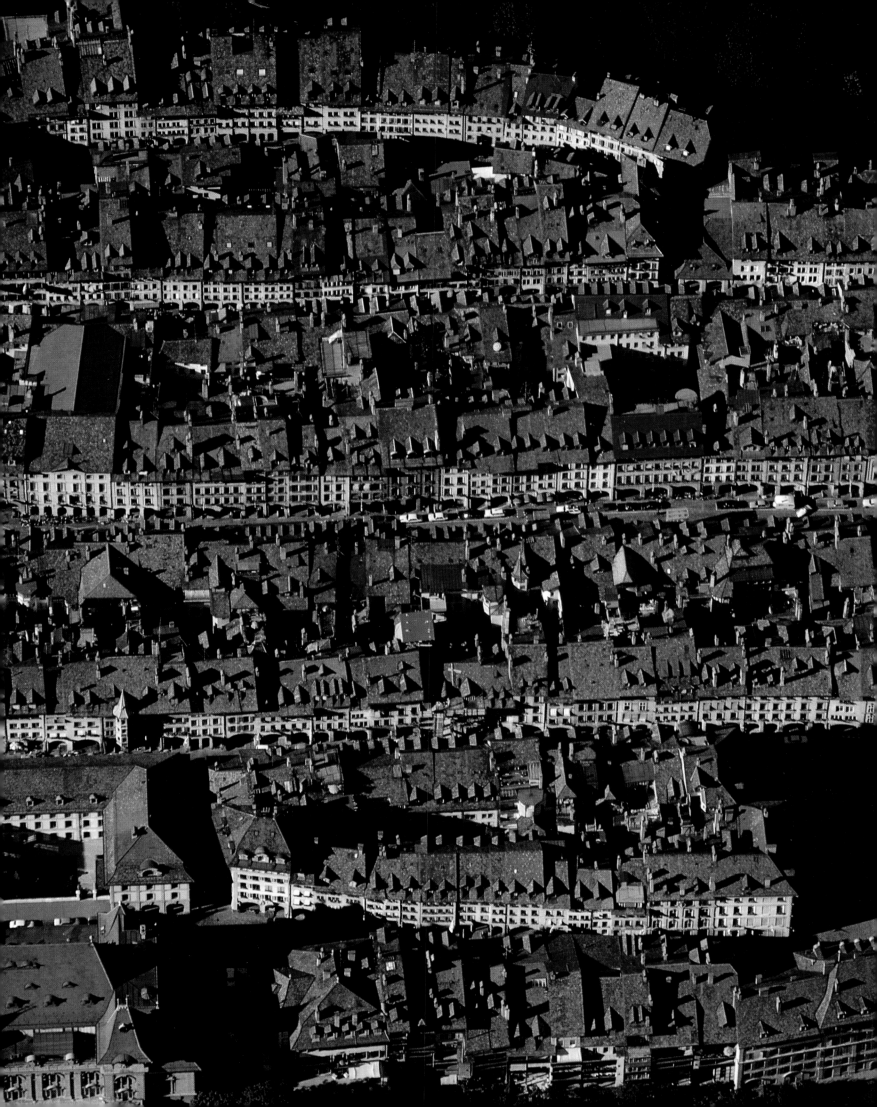

36-39

43

44 45
46 ▶

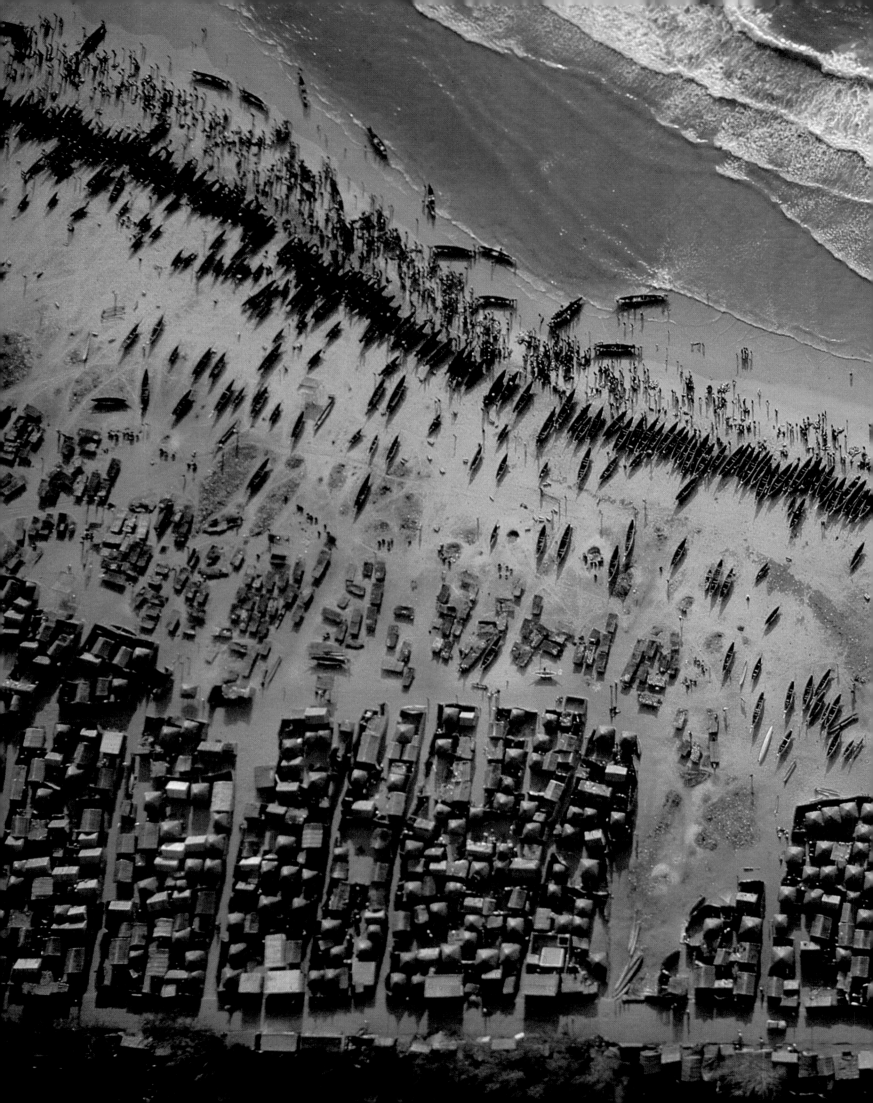

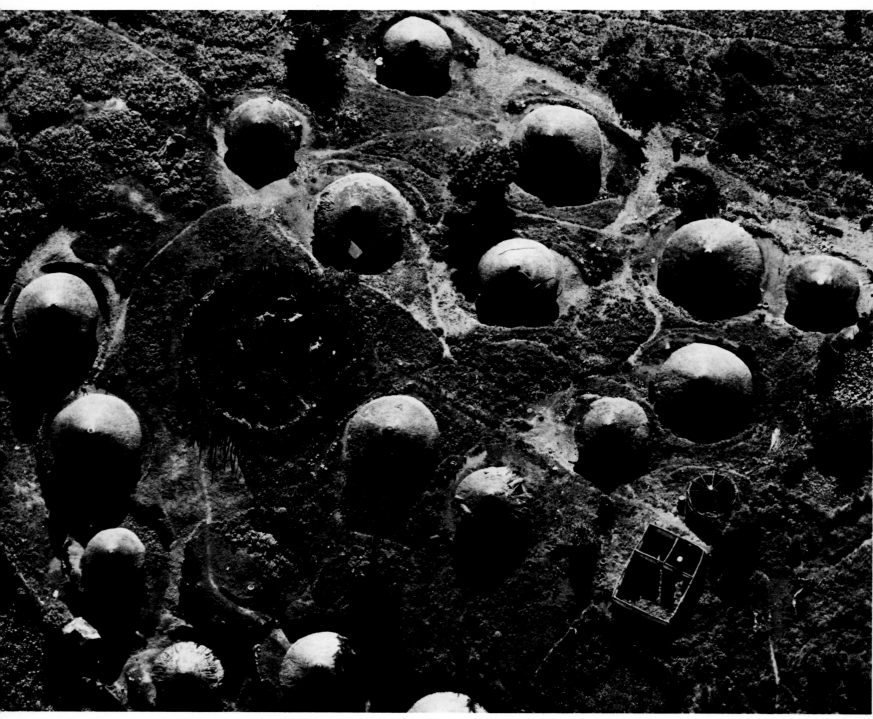

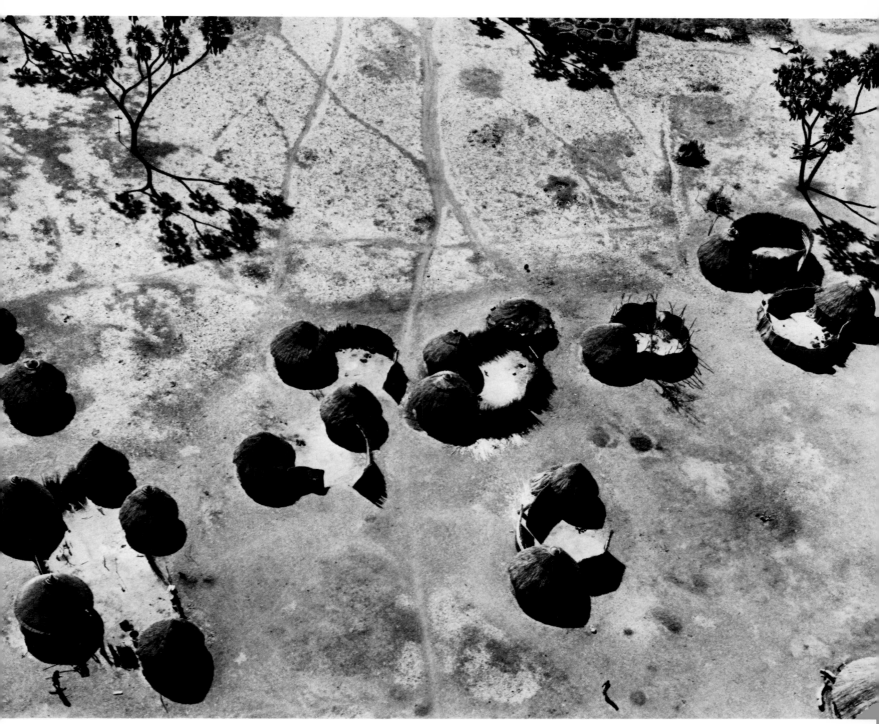

An archetypal settlement: the round town

An avalanche of urbanization now sweeps over the earth. It is anticipated that the urban population of Africa alone will have risen from 58 million in 1960 to 294 million by the turn of the millennium, that of Asia from 559 to 3,444 million, and that of Latin America from 144 to 650 million. The urban agglomerations are growing twice as fast as the population of the earth, and by the year 2000 more people will be living in towns than in the country. Uneasiness about the desolation of the towns is growing at the same rate. In our public discussions we are still mainly concerned with the problems caused by motorized traffic. Town planning, however, means more—or should mean more—than the mere sanctioning by the public of the private dream of moving on our own four wheels. And urbanism, reduced to the concept of supply and demand by the application of planning equations (man = dweller, man = eater, man = worker, man = commuter), becomes no more than a means of fulfilling physical functions.

Sibyl Moholy-Nagy, in her influential book *Matrix of Man, An Illustrated History of Urban Environment*, convincingly urges us not to consider towns simply as villages that have come of age. A town, she says, like a village , is a prerequisite of environmental control and the organization of human relations. To prevent aimless urbanization by an unthinking humanity, models of successful towns should be heeded. No settlement is more ideally suited for this purpose, none more dignified and significant, than the circular town. And an aerial view allows a diagnosis of its development as no other medium can.

When the circle—in itself the symbol of harmony, fulfillment and perfection—is further divided by a crossroad, the ground plan

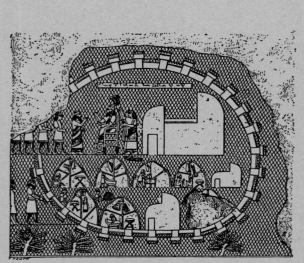

The camp of the Assyrian king Sennacherib (704–681 BC) on a bas-relief.

Fortified Assyrian camp as depicted in
Sennacherib's palace in Nineveh.

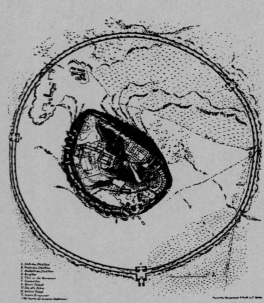

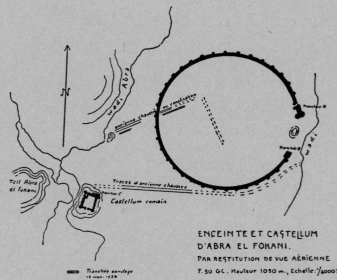

Zencirli, Aramaic period of architecture,
eighth century BC; diameter 2,300 feet.

Abra al-Fokani in Syria, late Assyrian
period (?), diameter 1,485 feet.
(All illustrations after *Creswell*.)

of the circular town is charged with archetypal meaning. The psychology of C. G. Jung interprets the quartered circle as a basic pattern of psychic self-knowledge. It appears in mystic visions, in dreams and myths, in signs of the Cabbala and alchemy, in stupa architecture (see Plate 172), in the mandalas and generally in the vocabulary of religious symbolism — wherever Man visually expresses the process of his self-fulfillment or his longing for spiritual wholeness.

Of course, circular town plans can also be justified by considerations of pure expediency. The shortest possible wall protects the largest possible area. Citizens whose houses stand on the same concentric circle all have the same distance to go to reach the temple in the center. And there are further benefits. The advantage of overall control of the whole center was temporarily applied even to prisons: the *panopticon*, first adopted in Belgium, in which one guard could overlook all the cells from a central point, so deeply impressed Thomas Jefferson (then the American diplomatic agent in Paris) that he sent news of this invention across the Atlantic. Yet however one turns it, considerations of this or any other utilitarian kind offend against the profundity of the original idea. The circular town saw itself as a likeness of the earth limited by the horizon — and as such it has always remained cosmically anchored: its center has always been the navel of the universe, and a hill or temple upraised there the umbilical cord.

The significance of the idea of the quartered circular town can hardly be measured by the number of times it has actually been carried out. The Egyptian hieroglyphic for "town," which also appeared after place names, was a crossed circle (as in drawings of ancient enclosures similar to the kraals that are still built in Africa today). Despite this, rectangular town plans by far exceed circular ones in the archaeological remains of the Nile region. The idea of the circle was evidently often accepted in the foundation ritual, but was rarely adopted when the town was actually founded.

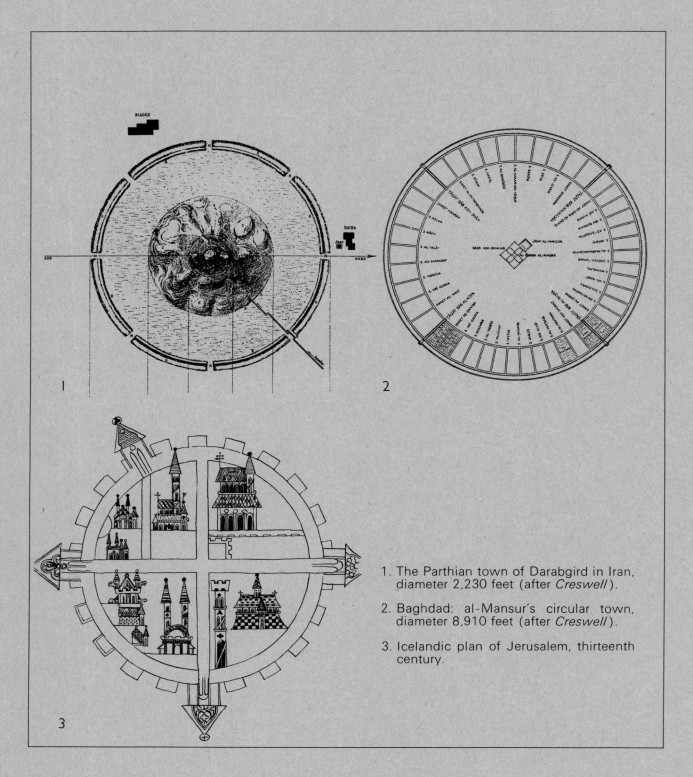

1. The Parthian town of Darabgird in Iran, diameter 2,230 feet (after *Creswell*).

2. Baghdad: al-Mansur's circular town, diameter 8,910 feet (after *Creswell*).

3. Icelandic plan of Jerusalem, thirteenth century.

Romulus, so Plutarch reports in his biography, commissioned Etruscan town-planning specialists to advise him on the foundation of Rome. On their instructions a furrow was plowed around the *mundus* (universe) or sacrificial pit to mark the confines of the town. In reality, however, ancient Rome was hardly a round town, and even Plutarch speaks, almost in the same breath, of a *Roma quadrata*, a square Rome.

The idea of a circular town, nourished by cosmic laws and sacred geometry, occasionally imposed itself even where observation contradicted it. Herodotus described the capital of the Medes, Ecbatana (founded in the seventh century BC), as a round town consisting of seven concentric walled circles, of which each inner wall overtopped the one outside it by the height of its battlements. The parapets of five outer walls were each in the color of a planet — white for Jupiter, black for Mercury, purple for Saturn, blue for Venus and orange for Mars, in that order from the outside. The two innermost walls, enclosing the king's quarters and the treasure house, shone in silver and gold — Moon and Sun. Herodotus' description was long accepted, and even Plato harked back to it. But neither subsequent visitors such as Polybius nor, later, the archaeologists were able to confirm the account given by the "father of history." At no time had Ecbatana been enclosed by walls — Herodotus had fallen for a myth. Jerusalem is a second example, closer to us in time, of an idea successfully defying reality. In Europe, by the Middle Ages at the latest, there had grown up the conception of a quartered world with Jerusalem at the center. This theory was followed in numerous literary and cartographic presentations which always portrayed Jerusalem as a circular town with crossroads. Was this meant to be a true picture? Every crusader must have known better (Plate 157). Or did these accounts perhaps refer to the heavenly Jerusalem? If so, they were in contradiction to the Revelation of St. John, which clearly describes the heavenly Jerusalem as square. Yet the psychically powerful

101

concept of the circular ideal town swept aside all the objections of Baedeker and the Bible.

If we accept Jung's theory of archetypes, the appearance of quartered circular towns — or the notion of them — in different continents and at totally different times in history raises few problems: wherever Man lives, he undergoes individuation and takes upon himself the cross of self-awareness. Archaeologists and town researchers, however, who think more about historical continuity, are perplexed by the distribution of the examples in space and time. The oldest round settlements so far discovered date back to Egyptian times. The round town also flourished on Iranian territory in the first pre-Christian millennium. But where do two typical examples of circular towns, the Central American Mexcaltitán and the northern European Trelleborg, fit into the pedigree? In the case of Mexcaltitán the discussion has not even begun. In attempting to explain the strict quadripartition of the Viking camp of Trelleborg, Werner Müller (in his book *Die heilige Stadt*) has groped his way far back along the hidden roots of Germanic and Celtic thought and custom. Where will the search for the origin of this urban pattern eventually end? Will it lead to a single primary find exactly fixed by the coordinates of space and time, or will it point rather to a number of isolated and spontaneous origins?

The observation that the numbers of concentric town plans increase enormously in times of political insecurity and change provides food for thought. Sibyl Moholy-Nagy mentions "an epidemic of cosmological town symbolism" in the Renaissance, and similar phenomena marked the end of the Ancien Régime and the French Revolution: in the architectural utopias of those times, circles and globes enjoyed striking popularity. And today? European futurologists met in the fall of 1972 in the French town of Arc-et-Senans (conceived as a circular town and an ideal industrial settlement for salt production by Claude Nicolas Ledoux just before the outbreak of the French Revolution, but only partly built).

And here they reinvented the round town. Concentric town plans — whether they are realized or remain blueprints is of secondary importance — are the flying saucers of urbanism: rotating images of salvation for the thirsting, overtaxed soul.

A Bavarian building law once forbade offenses against "symmetry and decency." Several authors of urban utopias, having considered the ethics of their ground plans, seriously believed that the linear town was a breeding place for vice and the round town a cradle of virtue. One may well ask: is life different in a quartered navel town? Is it different, I mean, for the ordinary citizen living on one of the outer concentric circles? (Not for one domiciled at the center as "Lord of the Four World Quadrants," "Axis of the World" or "Pole of the World" — all three actual Iranian royal titles.) Town planners do not yet have all the answers. The influence that the inner content and outer shape of a town has on its inhabitants is not yet understood. Will round-town people, aware of their privileges, ever destroy the shape of their city and voluntarily exile themselves from the center of creation? The fishermen of Mexcaltitán are still convinced of the cosmic significance of their settlement; for centuries they have only accepted inward growth by the renewal of quarters falling into decay. It is admittedly difficult to judge whether the encircling grip of symbolism has prevented the outward expansion of their settlement, or whether pure topographical necessity due to the confined situation in the lagoon has preserved the symbolic shape over long periods of time.

Other round towns, though cosmically anchored in the same way, have not been so successful. In 762–63 Caliph al-Mansur founded Baghdad as a circular town with four gates, but only a few centuries later there remained no trace of its original structure: haphazard building had erased the original circle. Much the same thing happened to Corona in California. It had been founded by one R. B. Taylor of Iowa, probably inspired by the concentric town utopias of the nineteenth century, in the form of a wagon wheel. The town hall and the meeting place formed the hub; from the center ran roads, like spokes, toward a ring road three miles long which cut the urban area out of the surrounding lemon, grapefruit and orange plantations. Speedsters, however, soon discovered a very worldly use for this cosmic circle: in 1913 Corona organized the first car race on its ring road, and the town was temporarily promoted to the "Indianapolis of the West." After this, little was needed to destroy the town's geometrical image completely; it overflowed into the landscape, and today only its name and the circular Grand Boulevard hint that Mr. Taylor had finer and nobler things in mind when he founded the town.

The center is the father of the circle, and the use to which the round town puts its center is an indication of the cosmic tension that holds — or may *not* hold — the town together. Auroville, a town now being built near Pondicherry, India, with a plan resembling a spiral nebula, has at its center a place of meditation inside a gigantic globe covered with gilded scales. In contrast, Sun City in Arizona is merely a mandala of the consumer society: here the focal point is a supermarket. Sometimes the center is politically determined. Ledoux put the "director's house" at the center of his ideal salt town, flanked by the salt boiling and filling buildings — monuments of the working world. Planners of the circular workers' settlement of Jupiá in Brazil originally intended to put a farm at its middle. Diagnosis of the center confirms the demythologizing of the round town in the modern world, but also reveals new gods. In the circular settlements near Copenhagen, which are aesthetically so attractive from above, the place where the axis of the earth used to be, or the holy mountain, or at the very least a temple, has been reserved for a parking lot for our modern four-wheeled Gog and Magog of the road.

58

The permanent Viking camp of Trelleborg near Slagelse on Zealand, Denmark. A circular rampart—of 440 feet inner and 558 feet outer diameter—encloses the main fortress. The circle deviates from geometric perfection by about one inch. Although this accuracy is relatively easy to achieve by rotating a rope around a central stake, the interior layout of the fortress proves that masters in the handling of the surveyor's rod and in marking-out techniques were at work here. A cross corresponding to the four gates divides up the inner circle. In each quadrant four elliptical boat-shaped buildings surrounded a square courtyard. The circular encampment possessed an outer fortress whose buildings, also boat-shaped, were arranged radially. When Trelleborg was built, at the beginning of this millennium, it had direct access to the sea and was presumably garrisoned by an army corps. Under Sweyn Fork-Beard (986–1014) the Vikings later built more of these round defense works with troop accommodation. The mathematical accuracy they demanded of their layouts reveals something of the spirit of the Viking military instructors in the barracks square. The ground plan of these circular fortifications might be regarded as the very ideogram of the quadri-partitioned circular city.

59

Ruins of the desert city of Hatra in Northern Mesopotamia, Iraq. The city had four gates; in a large square at its center stood palace and temple. Surrounded by walls and towers, the city flourished and became famous as a fortified settlement on the Roman—Parthian border and as a center of the caravan trade in the first centuries AD. An Aramaic dynasty ruled in Hatra under Parthian suzerainty. Among the beleaguerers it successfully defied were the Roman emperors Trajan and Septimius Severus. Hatra was abandoned in the fourth century. Its ground plan, though not geometrically perfect (with a diameter of approximately 6,550 feet), was probably an imitation of circular Iranian settlements. The Iraqi archaeological administration is at present rebuilding Hatra as far as possible.

60

Town and island of Mexcaltitán in the Mexican state of Nayarit. Mexcaltitán is situated in a lagoon on the Pacific coast, 155 miles northwest of Mexico's second largest town, Guadalajara. The shallow lagoon is alive with crabs, shrimps and fish, which are the population's main source of food and income. The settlement of the island dates back far into pre-colonial, pre-Spanish times. Some researchers even tend to equate Mexcaltitán with Aztlan, the original home of the Aztecs before they moved to their historical territories in the high valley of Mexico. Aztec legends about the origin and wandering of the tribe tell of an island in the middle of an inland lake, a paradise for fishermen and hunters of waterfowl. It is a fact that the memory of moon worship, Central America's oldest cult, lives on tenaciously in this village, and that its inhabitants are still quite convinced of the mystical-mythical significance of the circular settlement plan: for them, Mexcaltitán is the center of the universe. The cross formed by the four main streets inside the ring road that encloses the town mirrors the division of the heavens into the "world's four corners."

61

Sun City near Phoenix, Arizona, U.S.A. The Webb Development Company has been building its sun cities, since the success of the first near Phoenix, exclusively for the retired. No babies are born in these towns; at least one member of the family has to be fifty, and children still at school are only tolerated as visitors. These settlements for older people are built in flagrant violation of the prevailing opinion that old people abhor ghettos for the aged, and they are growing with youthful abandon despite (or perhaps because of) this geriatric heresy. Sun City near Phoenix has more than 23,000 inhabitants. When its architects selected the circle as the basic pattern of the town, cosmic symbolism was far from their thoughts. The curved streets serve to slow down the residential traffic and, more important still, the playgrounds of the inhabitants, the golf courses surrounding the city, can easily be reached on foot by every "sun citizen."

62

A Peul settlement on the Bani River, not far from its confluence with the Niger, Mali. The main streets run radially toward the central square with the mosque. The slight curve of these radial streets seems to set the whole pattern in rotation like a Catherine wheel. Can a place so vibrant with mythical symbolism still be called a village?

63

The foundation walls of a pueblo in the gorge of the Rio de los Frijoles ("Bean River"), New Mexico, U.S.A. The so-called Tyuonyi ruin in Bandelier National Monument is one of the most important Pueblo relics in the catchment area of the Rio Grande. Buildings of up to 3 stories with over 400 rooms lay in concentric circles around the central square with the *kiva*, the underground cult and assembly room of the Pueblo Indians. The ground-floor rooms were used as stores; only the upper

stories could be used for living, for the Indians had not solved the problem of a smoke outlet for their fire. In the east the circle had a gap to allow access; but the central square could also be reached via ladders over the roofs. The dating of wooden building material by dendrochronological methods points to a time between 1383 and 1466 — evidently the period of greatest building activity. Around the middle of the sixteenth century the Indians abandoned their homes in the canyon. Overpopulation, drought or disease — one or the other, or even all three — forced them to move on in the direction of the Rio Grande.

64

Wells Estate at Epsom, Surrey, England. The round area on which the estate has been built was the site of the mineral springs that made Epsom famous as a spa. (The only vestige of the fame of the watering-place today is the name "Epsom salts" for bitter cathartic salts.) Wells House, which exploited the medicinal springs, stood on a circular piece of land surrounded by Epsom Common. The private development which began after the end of the spa era accepted the circular boundary as the basis of its planning.

65

The settlement of Jupiá on the upper reaches of the Paranà in the Brazilian state of Mato Grosso. It was built in the 1960's for the workers who erected the Jupiá dam and power station (the first stage in the construction of the Urubupungà power station complex). The original intention was to place a model farm at the center of this trailblazing *vila piloto* for the working population — a programmatic pointer to an agricultural origin. But these ideological dreams did not mature, and even the plans for keeping the settlement alive after the termination of construction work finally petered out. The authorities of Mato Grosso refused the gift because its upkeep would have been too costly. The army took over one half of the circle, but the other half was delivered over to the bulldozer. The concentric town plan has indigenous antecedents: several tribes of the Amazonas Indians live in villages whose ground plans resemble huge spoked wheels, the huts being situated at the outer end of the spokes, on the wheel rim.

66

Nicosia, capital of Cyprus for more than a thousand years, has refused to be constricted by the circular wall erected by the Venetians in the sixteenth century in anticipation of a Turkish attack and has spread far beyond its fortifications. But the parts of the city *intra muros* are still Nicosia's heart and soul. The engineer who, between 1567 and 1570, sought to make the city impregnable sacrificed countless palaces, churches and monasteries to the encircling wall and the eleven bastions that projected into the city moat to permit the foe to be harried from the flank. But all in vain: after only seven weeks of siege, the Turkish troops took the town by storm in 1570 and tore down the cross from the cathedral. Thus the Gothic Hagia Sophia at the center of the city was converted into the Selimiye Mosque.

67

Summer houses in Brønbyvester, Denmark. The summer and weekend colonies near Copenhagen appear like clusters of blossom on the stem of the highway. The center of every cluster is the parking lot, a direct reversal of the situation in the classic wagon-protected laager—the people in their houses and gardens now protect their conveyances. The local council has leased the land for the construction of these settlements to a Danish allotment house society. After a maximum of thirty years the tenancy expires and the area must be returned in good order to the community. In summer the houses may be used continuously, but during the winter only at weekends.

58

59

66

Calligraphy of the industrial age

The benefits industrial Man derives from the aerial view steadily increase. Airplanes and satellites help in the preparation of plans for engineering projects — dams, roads, pipelines, railroads, etc. The plotting of maps and charts would be unthinkable today without aerial collaboration, as would the stocktaking of field and forest, area planning, the search for raw materials, the preservation of the environment, the gathering of up-to-date information on the state of our planet, the movements of sand in the deserts and of ice in polar regions, the variations of the snow line, the sediment transportation of large streams, the migration of cloud banks, and so on. Remote sensing not only uses the window that is open to us in the visible part of the electromagnetic spectrum, it also "sees" in the infrared and radar ranges. These Argus eyes in the sky, first used for military purposes, now carry out their espionage for Man. The information they provide is the preliminary for global management. This should not be forgotten.

The following pictures are a study of industrially active Man himself and of his appendix, the man of leisure, the consumer of spare time. The examples selected are calligraphies of our technical civilization, but they are more than just aesthetically interesting. Some undoubtedly belong to the controversial images of a changing era, and have triggered some widely differing reactions. The road through primeval forest, for instance, is certainly a manifesto — but of what? Only a short while ago the heart beat faster at the thought of pioneers advancing through the jungle: now it remains unmoved. Or the concrete lianas of the Pacific metropolis of Los Angeles — a great engineering achievement, but to what end? The most traffic-oriented of all cities was for decades a monument to

117

the internal combustion engine, symbol of a great conquest, until suddenly its pollution became notorious and even bumper stickers began to celebrate the days when sex was dirty but the air was clean.

68

A road through the jungle in the Brazilian state of Mato Grosso. President Juscelino Kubitschek, founder of Brasilia, complained about the lack of pioneering spirit in his compatriots who, he said, clung like crabs to the overpopulated shores of their huge country. He called up the spirit of the *bandeirantes*—those forest rangers, treasure hunters and adventurers who, in the seventeenth and eighteenth centuries, set out from São Paulo and penetrated the unexplored interior. "We must march west, turn our backs on the sea and stop staring at the ocean as if we were permanently thinking of sailing away." Kubitschek dreamed of Brasilia as the focal point of a network of roads which were to lure the Brazilians away from the coast into the enormous wilderness beyond. The tempo and spirit with which reality is today catching up on and overtaking Kubitschek's vision are alarming, to put it mildly. A gigantic road system 9,300 miles long for the development and colonization of the Amazon basin and the Mato Grosso is either already built, is in the course of construction or has at least been fully planned. Hardly is the bulldozer epic of the Trans-amazonica from Recife to Rio Branco completed when interest turns to the 2,500-mile *Perimetral Norte* from the Atlantic to Colombia, which runs through the jungle north of the great river. The last primeval expanse on earth, the planet's richest reservoir of oxygen and fresh water, trembles under tires and caterpillar tracks. But the sin against ecology is not the whole story: the steamroller of civilization brutally crushes the forest-dwelling Indios. They are massacred by the white man's diseases; his graders and planers, sometimes mistaken for super-tapirs, drive them from their homes ever deeper into the jungle. His Colt and his knife permit no waste of time. Colonization degenerates into colonialism of the worst kind. Brazilian Indianists have been courageously condemning this secret genocide in the virgin forest for the past fifteen years. There are even utilitarian arguments for the preservation of these forest-dwellers. They have survived in this green hell of poisonous and stinging plants, of snakes, bloodsucking ticks and insects, and they have learned to live healthily in it—an achievement made possible by their own medical, botanical and pharmacological knowledge. Though otherwise backward, they could therefore very well make a cultural and even economically valuable contribution to the taming of the jungle. But this consideration of expediency has so far been ignored in the conquest of the Amazon region. The frankly conquistadorian attitude of the modern *bandeirantes* often turns the dream road into a nightmare.

69

Ways for human beings: the steps of the Piazza di Spagna in Rome. A masterpiece of early eighteenth-century architecture, they connect the Piazza with the Trinità dei Monti church on the Pincio. The flower vendors at the foot of the stairway are part of the traditional setting of square and steps. The flower children at the tops of the flights hawking their homemade wares —belts and buckles, rings and bangles— are a new accent.

70

Ways for automobiles: city freeways in Los Angeles. A model for the control of traffic in agglomerations? Or writing on the wall, warning us against choking our cities with their own streets? They lace the body of California's biggest village like a concrete corset. Nowhere else are freeways knotted in such a fashion. The understandable euphoria about the relatively smooth flow of traffic (it lasted almost thirty years) has now been followed by a hangover. The stacking of the freeways today appears in a new and ominous light. We begin to see what town planners regard as the magic broom of individual traffic: freeways and parking areas eat up the last remnants of the city. "L.A. is a great big freeway," mocks the refrain of a popular song. Encaged in their cars, the freeway users are comforted by soothing words from the heavens. Pretty girls, hovering like rush-hour angels in helicopters above the traffic jams, advise the motorists on the quickest route to their office or home on behalf of a local radio station. They recommend patience: "Darlings on the Hollywood freeway! Remember that excitement is bad for your heart. . . ." In spells of acute smog their concern for the internal organs of their darlings down below may admittedly take on a desperate note: "Stop breathing. It's much better for your lungs." The commuting darlings meanwhile converse with one another by car stickers. The bumper aphorisms of the commuters speak volumes: "Remember when sex was dirty but air was clean?" "Boycott products of New York." "Drive carefully, they're waiting for your heart." "America, love it or leave it." "America, change it or lose it." "Honk if you're horny." "I'm a virgin." "Jesus saves." "Love thy neighbor, but don't get caught."

71
Factory bays in the dock area of Nagasaki on the Japanese island of Kyushu. A manufacturer of marine paints applies his products to the sawtooth roofs near the water's edge to test their resistance to light, weathering and the corrosive effects of salty air.

72
An air conditioning installation on a high-rise building in Los Angeles. Central air conditioners on the roof supply the whole building with air through a network of distribution ducts.

73
The Klondike near Dawson City in the Canadian Yukon. The discovery of alluvial gold in the Klondike sparked off the gold rush of 1896. Whatever escaped the gold washers and nugget hunters of the old days is now extracted from the diminishing gold deposits by mining companies using huge floating dredges. The systematically churned waste shows where the dredges have passed.

74
Bingham Canyon Mine in Utah, U.S.A. This is the oldest copper mine in the world in which the ore is extracted by open mining, and the largest single mine anywhere. More copper has been extracted from it than from any other mine. It is also the biggest of man-made excavations (the excavated material of the Panama Canal would only fill one fifth of it). In short: it is the most spectacular ore mine in the world. It is 2.4 miles long and 2,600 feet deep. In winter snow falls on its rim when it is raining at the bottom. Ore trains and trucks removing the surrounding strata commute on the "benches" of this amphitheater, which are up to 50 feet high and 120 feet wide. The daily transport performance of the trains and lorries can exceed 455,000 tons of ore-bearing rock — another world record. Since 1904 three billion tons of rock has been removed from the mine and ten million tons of crude copper produced from the ore. The gigantic crater and the gargantuan loading and transport equipment are a result of the very low copper

content of the ore and the unfavorable ratio of ore to country rock. Only the extraction of huge amounts of ore and strictest rationalization in its handling compensate for these handicaps. Bingham Canyon Mine belongs to the Kennecott Copper Corporation.

75
Breakwater in the Bay of Ise in Honshu, Japan. Because of the length of the coastline of the Japanese islands — approximately 6,000 miles — breakwaters are, not unexpectedly, in great demand. The blocks used for the stabilization of the shores come in all shapes and sizes, often in the form of composite cubes. Shore protection is not even their only purpose. In many places concrete gardens of this kind are heaped up to create suitable nesting and spawning grounds for fish.

76
Land reclamation near Sha Tin, Crown Colony of Hong Kong. Dump trucks are building a tongue of land into a bay of the South Chinese Sea. On it apartment buildings and a horseracing track will be erected.

77
Land reclamation in South Africa: windbreak fences for the tailings of a gold mine near Johannesburg. In eighty years the South African gold industry has hauled five billion tons of rock — thirty-nine times the excavated material of the Suez Canal — from depths of hundreds and even thousands of feet and crushed it to powder for the extraction of gold. The gold industry is obliged by law to protect the tailings (slime and sand dumps) against erosion by wind and water. A special organization, the Vegetation Unit of the South African Chamber of Mines, is responsible for covering these tailings with green plants. This is no easy task, as the mud and sand are extremely fine and highly acid and contain no plant nutrients. A further difficulty is the steep slope of the spoil dumps. Before they can be covered with vegetation, they must be stabilized by a trellis of windbreaking fences made of reed. Stabilization and cultivation are done by hand, but the effort is well worth while: the dust clouds which previously made life miserable in the lee of the tips will soon be a thing of the past — and the green hills themselves will be available as sites for apartments, schools, sports grounds, open-air cinemas and airports.

78
Sculpture of the fairground: a switchback railroad near Denver, Colorado, U.S.A.

79
Volleyball and basketball players at a high school near Santa Barbara, California. One ball is just dropping into the basket.

80
Competitors in a cross-country ski marathon in the Engadine, Switzerland. The road between Sils Maria and Sils Baselgia had to be prepared for the passing of the racers; by March it is in part clear of snow. The reproach directed at several of these European "people's races" ("lots of people, little race") certainly does not apply to the Engadine event. Lots of people, yes — 9,504 competitors in 1975 — but also plenty of racing: the marathon test over 26 miles long and at 5,900 feet above sea is a strenuous affair. It is not for nothing that the organizers require a medical certificate or equivalent from every competitor. The age of the participants is not limited upward, and

a 77-year-old has already taken part. The record so far: 1 hour, 42 minutes, 44.1 seconds (August Broger, 1975). But in this event even the last arrival is a winner — in the private test of personal stamina and willpower.

81

International rowers' regatta on the Rotsee, near Lucerne, Switzerland. Final of the eights.

82

Mothballs for aircraft: B-52 bombers, C-124 transporters and RF-84 reconnaissance fighters (righthand edge of photograph, middle) at the U.S. Air Force base of Davis-Monthan near Tucson, Arizona. Battle-scarred or outmoded planes and spacecraft that have been eliminated from the flying inventory of the American armed forces and Coast Guard are taken over by the Military Aircraft Storage and Disposition Center (MASDC). Some of these planes are reactivated. Civil authorities may also occasionally look for something suitable among the six thousand planes of seventy different types. The forestry offices of several states use old military planes for fighting forest fires. To put these planes in the air, the MASDC takes spare parts from other craft of the same model. Empty shells and stock that cannot be disposed of are finally sold as scrap. The same things that attract retired people to Arizona are good for the planes: the climate, plenty of sun, low humidity. Another advantage is the acidfree and noncorroding soil.

83

Mothballs for warships: units of the destroyer class of the U.S. Reserve Fleet in San Diego, California. Box-shaped super-

structures forward and astern protect the gun carriages from the sea air. Before World War II, the Reserve Fleet was satisfied if ships taken out of service did not sink. Now it has an arsenal of plastic sprays, protective coatings, drying systems and moisture-absorbing materials at its disposal in the battle against decay, mildew and every form of corrosion. The mothball fleet is moored in seven anchoring berths along the East and West Coasts and in the Pacific. It comprises some 700 battleships, aircraft carriers, cruisers, destroyers and auxiliary vessels of all types.

84/85

"The noble juice" (Shah of Persia): production plants of the oil industry. A forest of oil-well towers in the Lake of Maracaibo, below which Venezuela's richest hydrocarbon deposits lie; and billowing flames on the oilfield of Zelten, the first large field to be found in Libya. The construction of a gas liquefying plant near the oil port of Marsa el-Brega on the Greater Syrtis has now stopped this blazing waste, and the gas separated from the crude oil does not have to be flared off.

86

Rafts of floating timber in British Columbia, Canada. In times of critical ecological thinking the eye cannot be the sole judge of such a picture. Admittedly, almost a third of the earth's surface is clothed with forests, and as yet hardly a third of this third is being exploited. But are we to be lulled by statistics? The destructive lumbering which went hand-in-hand with the conquest of the West in the U.S.A. last century is now undisputed. Only the intervention of the Federal government around the turn of the century saved the forests of the Rocky Mountains and the Pacific

Coast Ranges. Luckily, this incredible shortsightedness in the past has sharpened the consciousness of North Americans for the importance of the forests. It has thus helped spread awareness of the fact that a timber trade that does not equate felling rate and growth rate is sawing off the branch on which it sits. But what of other parts of the world—the immense wooded regions of Asia, Africa and Latin America? Here complete clearing of the tropical rain forests heralds an ecological disaster.

87

A supertanker under construction in Nagasaki, Japan. The "Onyx" lies in the slipway, awaiting launching in a few weeks. With a length of 1,050 feet, a beam of 177 feet and a height of 85 feet, she has a carrying capacity of 268,951 tons and belongs, in the jargon of tanker shipping, to the category of the Very Large Crude Carriers (VLCC). She is an "oilephant" but not a "mammoth of the sea" (Ultra Large Crude Carriers of over 350,000 tons deadweight). The gantry crane bears the emblem of the construction firm Mitsubishi; owner of the ship is the French Compagnie Navale des Pétroles. Fifty percent of the world's ocean-going tonnage is accounted for by tankers; in mid-1975 there were more than 500 VLCC and ULCC ships. The only limit to jumboism in shipbuilding has long been set by the size of ports, straits and other shipping channels. The latest advances in automation and remote control, loading and unloading equipment, navigational and supervisory aids, seemed to preclude the possibility of accidents resulting in oil losses. The impossible nevertheless promptly occurred: fires, explosions and huge patches of oil on the sea made a mockery of the planners' claims. Supertanker pre-

sumption has thus run aground. Even for a standardized, less euphoric development of supertanker shipping there are now, after economic setbacks, obstacles not shown on any nautical map. But the braking distances in this branch of industry are long. A VLCC giant traveling at a speed of only six knots (just under seven miles per hour) takes just under one mile to stop with brakes fully applied.

88

Main station, Zurich, Switzerland: marshaling yards for passenger coaches. The bridges supporting the overhead contact wires divide up the picture. In a 24-hour day, 3,500 passenger coaches are made up into 600 trains in the passenger station of Zurich.

89

Color-coded containers parked in the container port of Hong Kong. Containerization has revolutionized the handling of goods in less than a decade. The goods are stowed in loading units that can be placed on rolling stock and lifted by cranes. The containers do not go from port to port only, but from loading station to destination. Together with specialized transport, loading and unloading equipment they have had a rationalizing effect which also pleases the eye. At present the containers are available in two standard sizes, the smaller unit being twenty, the larger forty feet long. More than a million of these units are currently traveling on oceans, rails and roads. An optimistic estimate is that world transport will need half as many again within the next five years.

90

A kind of large-scale Paul Klee: bridge building site near Santa Barbara, California. Old carpets, which are thoroughly soaked at least once a day, keep the freshly concreted fairway moist. When concrete hardens, gravel and sand are bound to the cement, which undergoes a gradual chemical reaction. At this stage, water is needed and heat is given off. In order to obtain good concrete, the builder protects it during this phase from loss of water and rapid cooling. Direct solar radiation and drafts are particularly harmful and may cause shrinkage cracks. The system adopted in California, though picturesque, is today out of date. In Europe special mats — a Swiss development — have come into use. They retain the heat developed as the cement hardens and make it unnecessary to keep wetting the concrete. This new procedure is more economical but optically rather dull — the so-called Guritherm mats are at present only available in green.

69

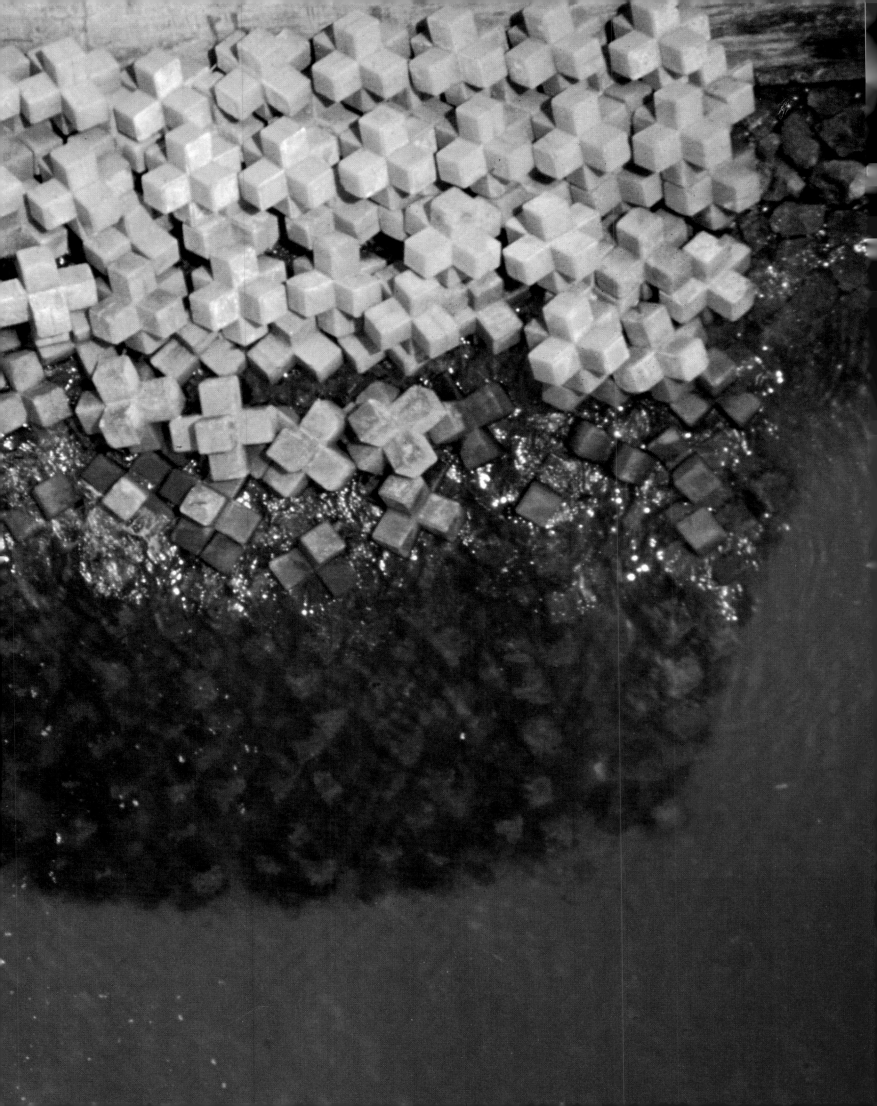

76

77

84

85

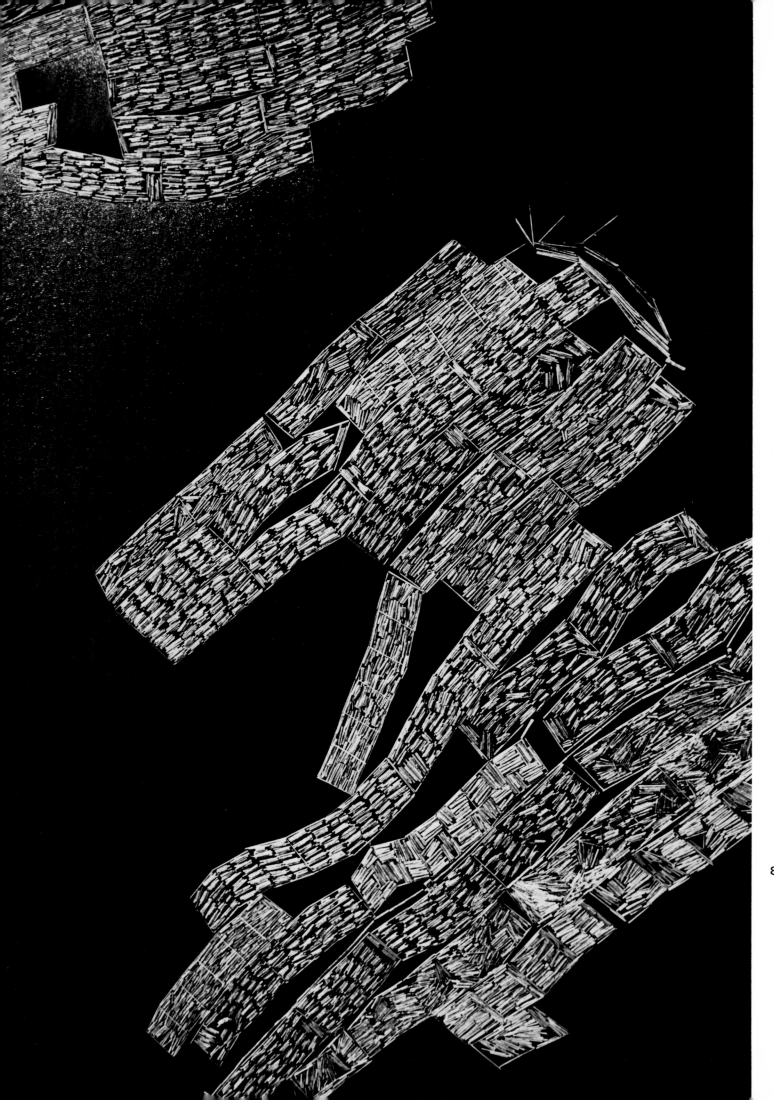

86

Saltworks as things of beauty

Kitchen salt, sodium chloride, is indispensable to human life. Contemporary Man, living in an abundance of salt, finds it hard to comprehend the Bible's "You are the salt of the earth" or the praise that other religions heap upon this life-preserving and decay-retarding substance. But he is at least not indifferent to *one* word that still reflects the ancient appreciation of salt, the word for the Roman soldier's "salt money"—*salarium*, our modern "salary."

The salt era began with the invention of pots and pans: cooking boiled away the natural salts contained in food. The hunger of humanity for salt has therefore existed ever since Man became sedentary and started to cultivate the land. Thus, the patterns of salt production have been with us at least since the New Stone Age. But the prismatic, polychrome effects that algae and other microorganisms produce in the gradually concentrated saline solution have only become visible since we have been able to look at them from above. And modern painting was needed to equip us to appreciate their beauty.

Sodium chloride is present on the earth in ample quantities. The oceans alone contain an amount of rock salt equivalent to almost fifteen times the mass of Europe (above the high-water mark). But these immense salt resources are unevenly distributed. Away from the coasts and from rock salt deposits, salt may be in short supply.

Power was guaranteed to those with access to rock salt or to salt that could be obtained from the sea or from brine springs, for it enabled them to satisfy the salt needs of others. From the Austrian Salzkammergut, possibly a center of the salt industry since Neolithic times, trade routes ran in all directions. Thanks to this salt trade, elements of the Hallstattan civilization in the Iron Age made their way far into Western Europe. But the powerful influence of salt is not reflected only in the faint pictures we can reconstruct of prehistoric times. Salines along the French Atlantic coast, another European salt supply center from Gallo-Roman times onward, were the cause of warfare between Breton kings in the ninth century. Control of the salt deposits in the Sahara was the mainstay of the African empires along the Sudan route, and the inhabitants of the Puebla highlands rebelled against the Aztecs' salt monopoly by joining forces with the Spaniards. Some of the ancient routes along which Homer's ''sacred salt'' traveled are still in use today (Plate 32).

The important part played by salt has been influenced from the outset by its preservative qualities. Salt makes possible (even in hot countries) a previously unthinkable storage of supplies, one of the prerequisites for urban civilization. In the adits of Hallstatt salt preserved remnants of the original Iron Age timbering, clothing and tools of the salt miners and even the corpses of workers killed in mine accidents. When discovered in the seventeenth and eighteenth centuries, these corpses were thought to be more or less contemporary victims and were buried—an irretrievable loss to prehistoric anthropology.

Popular beliefs and cults have always attributed life-giving powers to salt and have used it to ward off evil. Even science has recently discovered in salt something like a key to eternal youth: a German biologist isolated bacteria from 600-million-year-old beds of salt and brought them back to life in a nutritive solution.

91

The evaporation spiral *El Caracol* ("the snail"), near Mexico City. The soda works of Texcoco here concentrate brine from underground with the aid of sunlight. The brine, following a downward gradient, flows nineteen miles into the center of the spiral in a period of about six months, doubling its density on the way. From the center the solution is pumped into the factory at the edge of the spiral, where soda and salt are obtained from it. The owners are now testing the possibility of using the spiral for the production of protein. Experiments are being conducted in the outer whorls with salt-loving green algae whose protein content (approximately 65 percent) and possible yields per hectare of up to 45 tons in industrial cultivation recommend them as a source of food. Personal initiative is already exploiting this spiral for food production: the salt workers place twigs in sectors of suitable salt content, and a water bug lays its eggs on the twigs. From time to time they are collected and the eggs shaken off. They are in great demand as a sort of poor man's caviar. . . . The "snail" lies in the region of Lake Texcoco, one of five lakes in the high valley of Mexico. At the time when the Spaniards arrived here, the lakes still swelled in the rainy season to form a single water surface, the "Lake of the Moon" of the Aztecs. Water engineering measures around the turn of the century, together with the water requirements of Mexico City, have changed the situation radically. In the dry season Lake Texcoco is now a desolate hot dust pan over which, around noon, small tornadoes play hide-and-seek. The summer rains temporarily turn the bottom of the lake into a swamp. The great evaporative power of the sun in this part of the valley and the fact that Lake Texcoco once formed the lowest point of the basin explain the rich deposits of salt in the area.

92

Salt harvest near San Diego, California. The harvester — a caterpillar power shovel — removes the layers of salt in the crystallizing ponds and loads the trucks. Before use, the salt is washed and dried. That intended for human consumption is recrystallized and further purified. The crystallizing ponds supply ninety tons of salt per year and hectare. But before the sea water reaches them, it has passed through concentration ponds ten to fifteen times as large, in which it remains for years.

93–96

Concentration and crystallization ponds of the industrial saltworks in the bay of San Francisco and near San Diego. The heat of the sun evaporates the sea water that has been pumped into flat ponds. Salts of different solubility crystallize out one after the other. The coloring caused by micro-organisms in the brine enables the salt maker to judge the concentration of the solution, and thus the degree of maturity of the salt. The salterns along the American Pacific coast produce over a million tons of common salt per year. Only a very small part of this is meant for the table. Most of it is used in the manufacture of freezing mixtures, for water softening, in the canning industry, for meat pickling, for the curing of leather or as a chemical raw material. The degree of industrialization of a country can be measured by its salt consumption.

97/98

Salt pans near Massaua, Ethiopia, and Foundiougne, Senegal. Even pre-industrial salt manufacture creates fascinating patterns. In Africa, both in the east and the west, the evaporation pans of small salterns appear from the air like the palettes of painters. The salt makers near Massaua allow sea water to flow into artificial hollows. Those of Foundiougne take advantage of the seasonal variations of the water level in the swampy delta of the Saloum to fill their pans with brackish water. The sun not only concentrates the brine but also bleaches the extracted salt to an immaculate white.

99/100

The *marais salants* or salt marshes of Guérande, France. Europe has no other saline landscape that could compete with the individuality, beauty and historical significance of the salterns along the French Atlantic coast. Very little is now left of the once continuous production area between the estuary of the Gironde in the south and the Gulf of Morbihan in the north. Only the works on the peninsula of Guérande in Southern Brittany still give us an idea of the functioning of the original *marais salants*. This scene appears to the uninitiated as a chaotic labyrinth of canals and tideways, dykes and dams, settling ponds and evaporation pans — beautiful to look at but seemingly devoid of any practical purpose. In fact, it is the result of at least a thousand years of experience in the steadily perfected art of salt making. High tides and the gentle gradient of the sea floor on the French Atlantic coast create a wide tidal zone of which the *paludiers* take skillful advantage. Sea water flows by gravity through an elaborate system of storage, evaporation and crystallization ponds. The water supply is regulated by sluices. At the highest and outermost point lies the storage basin, the *vasière*. This can only be filled by the spring tide, as normal high water does not reach its inlets. The lowest and innermost part of the system is formed by the *oeillets* ("little eyes"), crystallization beds, each measuring 860 square feet, from the bottom of which the *paludier* scrapes the salt with a flat-edged rake (Plate 99). The year of the Breton salt maker begins in spring. The ponds and beds have to be drained, and the walls and bottoms of the ponds, dykes and dams have to be weeded, cleaned, leveled, trodden and beaten. In June, after a pre-evaporation period lasting up to seven weeks, according to the hours of sunshine, the *paludier* harvests for the first time. Given favorable weather — plenty of sun and wind — the *oeillets* will go on producing without interruption every second or third day until far into September, each yielding up to three tons of salt. The yield, piled in bright heaps, is left to dry on the dams between the beds and on the rim of the evaporation ponds. Sometimes it even winters there, protected by a coat of reeds and soil. Precipitation is the salt maker's worst enemy. A summer storm dilutes the brine and delays the maturing of the salt by eight days, and a rainy summer can completely destroy any hope of a harvest. In the Middle Ages, French Atlantic salt acquired European importance. Danish kings waylaid Hanseatic, Prussian and Livonian vessels carrying the coveted cargo. Since the middle of last century, however, the position of the Breton salt makers on the market has gradually deteriorated, even in France. Modern railroads have helped to popularize the rock salt of Lorraine and the marine salt from the salines of Languedoc. It matters little that connoisseurs praise the off-white salt of Brittany because of its unique fragrance of violets! Competitors who were unable to copy the scent were at least able to imitate its gray color. Nothing now seems to be able to check the shrinkage process of the *marais salants*. Every year new *oeillets* go out of service; there is no stopping the advance of industrial salt production. At present, some 300 leaseholders or independent workers operate approximately 15,000 *oeillets*. In Saillé (the "salt capital") the *paludiers* have since 1971 made their appeal to tourists in a touching museum in which the craft, costumes and life of the salt makers are immortalized. Is this a sign that their time has come? A defense committee was set up about the same time with a new marketing organization for Breton salt; its object is to prevent a landscape of highly individual character and its inhabitants from becoming a piece of dying folklore.

91

93–96

97

98

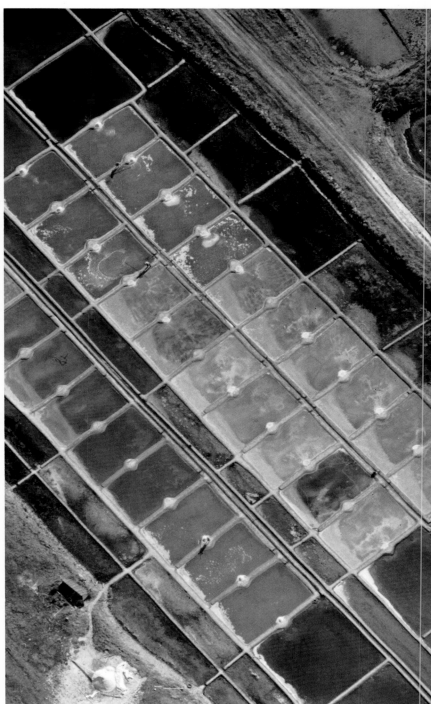

99

The farmer as artist

The farmer, living on the fertility of the soil, multiplies the beauties of the earth from above as no one else can. In their plastic, monumental effect some of his works surpass the most beautiful architectural creations. What is the Chinese Wall compared to the rice terraces of Banaue? The farmer draws with plow, harrow and combine harvester, paints with the colors of his crops. The shapes and hues of the agricultural world seem inexhaustible from the air. The following photographs are the foundation of a new science. In the absence of a more pompous name we might call it "fieldology": the comparative study of the fields of different peoples. There is more to this subject of research than might at first appear. Laws of inheritance, local land rulings, mechanical aids, land forms, rotation of crops—whichever factors influencing the structure of the fields are isolated (legal, historical, industrial, agrarian or socioeconomic), there almost always remain others that cannot be evaluated. It is as if the farmer had found the philosopher's stone, which he uses to turn ordinary earth into visual gold.

The artistic effect is automatic and runs through the cultivation of potatoes and clover alike. Since the 1960's representatives of a new and intentional landscape art, such as Trara, have appeared on the scene (Plates 176 and 177). Should the farmer's unintentional creations now be classed as Land Art before the act? Or are those professional artists who grub and burrow in meadow and desert merely latecomers, imitators? A comparison of these works —the casual and fortuitous as against the studied and laborious— leaves hardly any doubt as to who—literally and figuratively— holds the field.

100

161

101

A fire of straw in Zealand, Denmark. The patterns sketched by the combine harvester are soon being embroidered with the black threads left by fires; for if the fire is not whipped ahead by the breeze, it creeps along the rows of straw sheaves as if they were fuses. Though it devours the straw greedily, it hardly singes the stubble in between, and it does not harm the microorganisms living in the topsoil. Burning of the straw in the fields is relatively new. Today there is not enough labor to carry it away, and the demand for straw has in any case dropped because of the reduced number of horses on the farms and new stable systems requiring no litter. The only solution, therefore, is to burn it. The calcium and phosphorus of the stalks are retained in the ashes. Notification of the local fire brigade is prescribed by law, but losses are nevertheless unavoidable. "Yesterday," a Danish newspaper reported after the particularly hot day on which this photograph was taken, "yesterday we lost only one farm." Such equanimity is not universal, and this method of clearing the fields has its opponents.

102

Terraced rice fields in the north of Luzon, Philippines. The rice growers of South and East Asia are world champions in the art of terracing, the creation of flat strips along the hillsides where rice can be grown. The Ifugao, an old Malayan tribe that settled in the north of Luzon 2,000 years ago did not shrink (and still do not) from even the steepest slopes. They built terraces up to fifty feet high with rough scree material — and all to obtain a terrace only ten feet wide. The walls are made somewhat higher than the cultivated area and thus serve as dikes and footpaths. Water from springs and rivers is conducted to the fields in channels and bamboo pipes so as to make up any defiency in local precipitation. Rice harvesting takes place in June and July, the remainder of the year being given up to the cultivation of sweet potatoes. In the region of Banaue, the heartland of the Ifuago, the total length of the terraces is equal to the circumference of the earth. Many mountains are artificially terraced over a difference in altitude of over 3,000 feet. The upper limit of rice cultivation here lies around 5,500 feet — and the diligence of the Ifuago does not give a single foot away.

103/104

Terraces for the cultivation of grain, fruit and vegetables in the mountains of Judaea. These mountains are the roof of Palestine, and an outstanding example of the labor involved in God's command to Israel to subdue the earth. The toil that turns stones into bread also turns mountains into sculpture. Millions of terraces that have existed since Biblical times cover roughly half of the total area of the Judaean highlands. The terracing takes into account the run of the geological strata (hard, dark, angular limestones from the Upper Cretaceous), the angle of inclination of the slope, the amount of water available, solar radiation, and so on. Possibly the first terraces were built where a spring bubbled for at least a few months every year. Today dry farming predominates. Northern slopes are terraced more extensively because they get less sun and remain moister in winter, so that weathering leads to speedier soil formation. Settlements are located on crests and rounded summits. The inhabitants accept the disadvantages of such a position because it saves arable land that has been wrested from nature in the course of the millennia.

105

Terrace cultivation of millet in the Mandara Mountains, Cameroon. The terraces curve around the slopes like contour lines in a relief model. They reveal the pathos and intensity of a landscape created out of necessity by the Kirdi (Plate 38). Driven away from the fertile plains, they are forced to lead the life of hillmen on meager soil. Their terraces prevent the washing away of the topsoil. In the northern Cameroon highlands millet is grown in the rainy season and is harvested at the end of October.

106

Field of Californian poppies (*Eschscholtzia californica*) near Lompoc, California. The flowers are left to wilt in the fields, and machines collect the seedcases. Lompoc's long, rainless summers create ideal conditions for the harvest, and the valley meets more than half of the world's demand for seeds for the garden. The *eschscholtzia* was discovered by the poet and botanist Adelbert von Chamisso, who, between 1815 and 1818, sailed around the world with Otto von Kotzebue's Russian expedition. He named the flower for a fellow traveler, the zoologist J. F. Eschscholtz.

107

Terrace farming in the prefecture of Saga in Kyushu, Japan. Ricefields and tangerine plantations occupy the terraces. Kyushu, the most southwesterly of the main islands, is famous throughout Nippon for its tangerines.

108

Rice harvests on the Niger, Republic of Niger. The tribes living on the great bend of the Niger have for centuries practiced an archaic form of rice cultivation. They do not cross or improve the strains; they do not transplant, manure or weed; they do not protect the rice against insect pests; and often they do not even sow it. The half-wild varieties of *Oryza glaberrima*, a red water rice that grows here, are almost a gift of the river. In December, when the rice is ready to harvest, the Niger is in spate downstream of Timbuktu, and the rice has to be cut with a sickle from a canoe. To protect the stalks and panicles — the paddy — from wild animals, they are temporarily stored in thorn enclosures on the banks. Red rice, probably indigenous to the Niger, is undemanding, but the yield is meager: at best about 1,500 pounds of paddy per hectare.

109/110

The oasis landscape of the Souf in the Algerian Sahara — archetype of all oases. The ideas, emotions and expectations aroused by the word "oasis" are here combined in a single image. The huge dunes are like the waves of a waterless ocean. In a thousand craters that have been wrested from the sand and are now protected against it by rows of fences made of palm fronds, groups of date palms grow, their crowns often hidden behind the rim of the crater. Ten thousand cupolas and barrel vaults crown El Oued, the capital of the Souf, and its surrounding settlements and isolated farms (Plate 34). Since the water did not come to them, the Souf dwellers went with their palms to the water; they planted the trees in hollows of various shapes, sizes and depths so that the roots reach into seepage water beneath the sand. In other words, here is a victory over the desert. Up to two or three hundred palms are assembled in the larger hollows which in the south of the oasis region, may be up to 165 feet deep. The palm gardens of the Souf, which ride like ships on the waves of the dunes, are undoubtedly a paradigm of Man's successful stand against an unpropitious nature. But even here the desert has the last word. Oil finds—which have already brought traffic signals, one-way signs and parking meters to the oases—now threaten date palm cultivation. Today's motorbike-riding sons are fed up with their fathers' victories. The long distances to the palm gardens are too much for them, to say nothing of the drudgery in the hollows. Crater after crater is abandoned to the sand, the desert takes back for ever what an earlier generation wrung from it by cunning or defiance. In addition, modern palm plantations, irrigated from deep wells, are destroying the old gardens. Surplus water from the depths swells the groundwater stream under the surface; it rises in many places, and the palms are drowned.

111

Part of a *foggara* near In Salah in the Algerian Sahara. *Foggaras*—better known outside of North Africa under the name of *qanat*—are horizontal wells. Above ground they appear as rows of shaft openings, mostly in soldierly alignment but sometimes forming an untidy scrawl. Every opening wears a collar of excavated material. Whether arranged in military file or in haphazard meanders, the openings mark the underground course of a tunnel that runs from the source at the head of the system right through to its outlet into an open supply system (for the irrigation of a palm garden, for instance). The parent wells that tap the groundwater are up to 130 feet deep in the Sahara. In the Iranian plateau, the country of origin of the *qanat*, they may attain depths of 980 feet. The tunnels, through which the water flows by gravity, may be two, three, sometimes even

twenty-five miles long. At regular intervals, shafts lead down to the channels. They serve for ventilation, for the removal of the excavated material, and as manholes for maintenance work. The construction of a *foggara* 2.5 miles long at a mean depth of 40 feet, with ventilation shafts every 32 feet, requires 48,000 working days. A *foggara* also needs frequent servicing. The very gentle gradient of the stream—about one-hundredth of an inch per foot—makes it extremely sensitive to debris crumbling from the roof or to sand blown in through the shafts. Thus the watercourse is continuously exposed to the danger of thrombosis. Should the roof sink, the *foggara* has to be put in working order again by making it longer or deeper or by providing it with "feet" (side tunnels). The *qanat*, a method of obtaining water as desperate as it is ingenious, was invented in Persia more than 2,500 years ago. *Qanats* supplied water to Ekbatana, capital of the Medes, in the seventh century BC and, somewhat later, to Persepolis. The Persians, the Jews returning from Babylonian captivity, the Romans and the Arabs spread the *qanats* as far as China, in the whole of the Near East and in North Africa. From there the Arabs introduced them to Spain and the Spaniards brought them to the New World, to Chile and Peru. Even our word "canal" contains a memory of them, for it is cognate with the Semitic *qanat*. In modern Iran drinking and irrigation water is supplied by some 40,000 *qanats* with a total length of about 100,000 miles. In contrast, the *foggaras* in the Sahara are dying. Using no other implement than a short-handled hoe and a basket, the *foggara* builders formerly burrowed into the depths and with the same primitive tools dug a tunnel from the bottom of one shaft in the direction of the next. Hundreds of workers must have lost their lives, trapped and buried in the shafts, suffocated or exhausted. The human moles that made the *foggara* were slaves, and the arrival of the French in the great desert put an end to slavery and thus to *foggara* building. Since the turn of the century, few new *foggaras* have materialized. Of the old ones, one after the other caves in for lack of maintenance or is covered by advancing dunes. The total length of the *foggara* systems in the Sahara is presently estimated at approximately 1,850 miles.

112

A water engineering trick for increasing the amount of runoff from slopes: old walls of loose stones and heaps of flint in the Wadi Abiad near Shivta, Israel. Their continued existence is endangered by the use of the Wadi Abiad as a training ground for tank maneuvers. The mysterious heaps of flint—sometimes up to 600 per hectare of slope area, all in all hundreds of thousands—are spread around half a dozen ruined towns in the Negev. They were already noticed by travelers in Palestine last century. The Bedouins call them *tuleilat al-anab* ("knolls for the vine"). The first attempts at solving the riddle of their origin were misled by this designation. Neither the heaps nor the dams of surface rubble are directly connected with viticulture. Did they promote the washing down of fertile soil into the cultivated valley—a form of controlled erosion? Or was their purpose to improve the local climate, particularly to stimulate dewfall? These questions today seem to be off target. Since 1954 a team of archaeologists, botanists, water engineers and agricultural specialists under the leadership of Michael Evenari have been groping toward a more plausible explanation. The surface stones, they now claim, were collected and simply piled up in heaps or rough walls in order to increase the amount of runoff from the slopes. When rain falls, it transforms the open soil within minutes into an almost impermeable mass over which the water, instead of seeping away under the cold stone cover, now runs down much faster into the cultivated valleys. The rubble dams are also helpful in guiding the water to pre-selected plots of land. Experiments have shown that the *tuleilat al-anab* may double the water runoff from the slopes or even triple it, after a short shower, which is the prevalent type of precipitation on the margin of the desert. With the help of this simple form of natural irrigation farmers formerly cultivated fields, vineyards and orchards here for eight hundred years, under the dominion of the Nabataeans, Romans and Byzantines, from about the middle of the third century BC— relying only on the meager rainfall and their own clever know-how. Evenari decided to check this finding in a practical test. Near Shivta (the ancient Subeita) and Avdat (the ancient Oboda), he provided two farms with all the old irrigational tricks— and lo and behold, apricots, plums and peaches, pomegranates, almonds and olives, carobs, pistachios and grapes were soon blooming and fruiting in the desert.

113/114

Cattle watering places in the Pampas, Argentina: a natural pool and a well operated by a windmill. The Pampas have little flowing water but are rich in hollows with no outlets. Admittedly these have a limited usefulness for stock farmers because many of them become so salty that their water is unfit even for cattle, while others dry out in years of low precipitation. The farmers therefore drilled thousands of wells for cattle, sheep and horses. Wind power raises the groundwater from various levels in open containers. The Pampas are the Argentine's economic heartland, both for cattle breeding and for tillage (Plate 119). The ocean of grass the Spaniards found around the basin of La Plata (it deeply disappointed them because they expected to find silver and gold) now lives on only in gaucho songs. Pampa, a word from the Quechua language, means roughly "treeless plain"—and since Darwin, geography has had a Pampas problem: Was the grassland at the time of the Spanish colonization a natural form of vegetation? Or had hunting Indians burned down the forest in pre-Spanish times? Research today favors the former assumption and declares Man not guilty of having caused the scarcity of trees in the Pampas at that time. In contrast, today's Pampas are completely Man's creation—including even trees offering shade to cattle and gauchos. There are also clear signs of overcultivation. Too many wells exhaust the groundwater supplies near the surface; and the replacement of the natural plant cover by farm crops, particularly the demanding alfalfa, further increases the deficit in the water balance.

115

Remains of Aztec agriculture near Mexico City—the *chinampas*. The high-lying valley of Mexico has always been poor in land suitable for intensive cultivation. The swampy shores of the five lakes in the mountain-fringed basin and the marshes between them challenged the inventive spirit of the peoples who settled here as much as two thousand years ago. They replied with exemplary methods of drainage. By digging out the swampy ground and transferring it to rectangular enclosures

made with wickerwork, they created artificial islands—the *chinampas*. Water flowed into the trenches and transformed them into canals. They planted willow trees around the edges of the *chinampas*, and the roots eventually replaced the artificial enclosures. The Aztecs—late arrivals in the valley—at first had to be content with a few islets in Lake Texcoco. By applying the *chinampa* technique, however, they gradually transformed the area into the glimmering capital the Spaniards later admired as a second Venice; and the labyrinth of the *chinampas* in the shallow lakes produced the agricultural plenty which was a prerequisite for the military and political expansion of the new overlords. The remains of the *chinampa* islands on the periphery and in the surroundings of Mexico City today supply the capital with flowers and vegetables. The photograph shows a piece of *chinampa* landscape on the edge of the old Lake Chalco. The boundary of the natural *tierra firma* is in many places as straight as if it had been drawn with a ruler. The orientation of the grid created by the long, narrow islands corresponds to that of the street system of the ruined city of Teotihuacan. The *chinamperos* themselves are a fairly pure remnant of the Aztec people and speak only Aztec (Nahuatl) among themselves.

116

The star of Ensérune in the Département of Hérault, France. In 1247 three feudal lords from the region of Montady and the notary of Béziers obtained permission from the archbishop in Narbonne—he was responsible for water resources—to drain the Etang de Montady and to make arable land out of it. The rays of the pattern which resulted from the reclamation work are drainage ditches. The optical and technical aspects of the pattern thus contradict each other: the star does not radiate, but collects. The drainage waters flow down the gentle gradient of the ditches into a collecting sump at the center; from there, a canal that cuts through the hill of Ensérune in an underground conduit leads the water into a neighboring pool. Bylaws outlining the rights and duties of the owners of the reclaimed land have remained absolutely unchanged since the thirteenth century. The *Syndicat Agricole* of Montady/Ensérune is the oldest in France and one of the oldest in Europe. The whole reclaimed area is slightly oval, a consequence of the terrain. The individual sectors of the circle take this into account, becoming wider or narrower, but all the plots are of the same size. For centuries the more than five hundred hectares of the area were reserved almost exclusively for vine cultivation. Since 1956, however, when a sudden onset of cold following an Indian summer caused severe damage, the farmers have tried diversification. Fruit tree, tomatoes, maize, sorgo and lucerne grow on the sectors today.

117/118

Vegetable growing in Imperial Valley, California, and in the New Territories, Hong Kong. Two types of fields, and two philosophies of life! In the first we see the dictate of the straight line, cultivation as subjugation, the conquest of nature by geometry. Imperial Valley as a whole is land wrested from the desert, a triumph of irrigation technology. In the second example: the Taoist philosophy of respecting and adapting oneself to the landscape, a tacit understanding with nature, agriculture as a way to cosmic harmony. Geomancy—*fêng-shui* ("wind and water")—

ensures that the homes of the living and the dead are in accord with the breath of the world that can be felt flowing through every landscape. A geomancer used to be called upon when a Chinese farmer wanted to fix the boundaries of new fields. The earth diviner rejected the straight line as soulless, and geometry as godless; in his opinion paths and watercourses should wind, fitting naturally into the landscape, for adaptation, not resistance, is the guarantor of good fortune.

119

Summer harvest pattern in the Pampas of Argentina. The Pampas farmers are spontaneous draftsmen and painters. Their materials: the clover, wheat, maize and oilseeds they grow in their fields; their tools: the agricultural machines. And their compositions furnish food for thought—for art enthusiasts and museum directors. The beauty of the field designs, moreover, is ephemeral. Or should the farmer be offered a subsidy to let his combine harvester stand and rust at the point beyond which he can only destroy his composition?

120

A field quadrangle in the Brazilian state of Mato Grosso. The pattern is formed unwittingly by agricultural work in the *terra roxa*, where the gneiss foundation of the country disappears, toward the Rio Paranà, under volcanic strata. The farmer is plowing green plants under; the graduation of the red shades reflects the stages of his work, the lighter areas being drier. It is the morning of the fourth day, and the plower begins the last phase of his work.

121

A field mosaic in the province of Wollo, Ethiopia. After the rains, the *nug* fields are

167

still bright on the tableland while the farmer begins to turn over the soil for·a second harvest. *Nug* (*Guizotia abyssinica*), a member of the sunflower family, is Ethiopia's most important oil plant. Its rich seeds yield a valued edible oil. Ethiopia is the only country in tropical Africa that used the plow as well as the mattock even in precolonial times—part of the cultural heritage of settlers from Southern Arabia.

122

Citrus plantations near Morphou, Cyprus. The citrus orchards of the Mesaoria, the fertile transverse depression between the two mountain ranges of the Mediterranean island, enrich the Briton's breakfast with grapefruit and oranges. The citrus industry in Cyprus is young, but it is already reaching its production limits. Morphou, the main citrus-growing area, suffers from a deficiency of water. The Mesaoria ("land between the mountains") lies in the rain shadow, and the groundwater reserves are quickly exhausted.

123

Coffee plantation in the Brazilian state of Saõ Paulo. The northwestern part of this state lies outside of the classic coffee-growing area. Here the loamy red earth in which the coffee shrub thrives is present only in patches among the unsuitable sandstone. Measures taken to preserve and improve the soil, combined with the shapes of the landscape, produce these curious patterns that look more like fingerprints than plantations.

124

Pineapple fields on the island of Oahu, Hawaii. The cultivation of pineapple on the Hawaiian archipelago follows a three-year cycle: the first harvest takes place in the second year, the second harvest in the third year, and then the plants are plowed under. The processing facilities—Hawaii produces more canned pineapple juice and slices than any other region—are only in operation for a few weeks in summer. The exact date of the harvest is therefore carefully planned. The pineapples ripen in eighteen to twenty-four months, according to whether the plants are grown from root suckers or from "crowns" (the leaf tops of the pseudocarp). Even more exact scheduling can be achieved by gassing the plants with ethylene, which terminates the growing period and induces fructification. The graphics of the plantations reflect the optimization of growing methods. Every field is twice as wide as the arm of the sprinkler truck (for spreading liquid manure and pesticides) and as the conveyor belt of the harvesting machine. The rounded corners of the fields match the turning radius of the vehicles. We hear a great deal today about the technical utilization of solar energy and often forget its photochemical use in nature. Fossil fuel, at present (and for a long time to come) our main source of energy, originates from the storing of solar energy by green plants —by far the most important chemical process on earth. Natural plant growth utilizes only about 0.1 percent of the sun's energy calculated in thermal units. Compared to this mean value, the perfected pineapple plantation is approximately thirty times more efficient as a "solar power plant"— the same is actually true of the virgin rain forest in the tropics. A further tenfold increase in the efficiency of photochemical storage of solar energy can only be attained by a different kind of agriculture: the utilization of green algae in cultures or suspensions under optimal conditions.

125

An oyster farm near Marennes in the Département of Charente-Maritime, France. The oyster fattening grounds (*claires*) lie in the estuary of the Seudre. Mixing of sea water and fresh water here creates the salt content that is most favorable for the oysters. The farms are mostly abandoned salterns (Plates 99 and 100). Oyster breeding along this part of the French Atlantic coast has made up for the losses caused by the decline of the salt industry. Oyster larvae drifting with the current settle on tiles, stones or wooden stakes that are scattered or planted in the tidal flats for this purpose. After two years at the most, the breeder collects the young oysters and fattens them in the *claires* until they are ready for the market. In France oysters are generally four to five years old when they are sold. Gourmets praise the Marennes oysters for their particularly fine aroma. This is due — as is their typical green coloring—to a microscopically small alga which is only found in the waters of Marennes. The Département of Charente-Maritime produces 600 to 700 million oysters a year.

126

Bamboo rafts with pearl oysters in the Bay of Ago, on the Japanese main island of Honshu. The view that a diamond is a girl's best friend has been shared by very few Japanese since Kokichi Mikimoto, the son of a poor noodle-maker founded a million-dollar industry with the breeding of the first spherical pearl shortly after the turn of the century. A tiny grain of mother-of-pearl from a freshwater shell is surgically sewn into the connective tissue of a pearl oyster; this artificially stimulates the growth of a pearl. The inoculated shell is placed in a wire basket which is suspended from a bamboo raft, and these rafts are a characteristic feature of the shallow, sheltered bays along the peninsula of Shima, center of the Japanese cultured pearl industry. The pearls mature within three to six years, according to water conditions. The baskets are brought to the surface from time to time to clear algae from the shells and check their health. Typhoon warnings, too, necessitate the salvaging and evacuation of the baskets and their inmates. When listing his worst enemies, no pearl breeder forgets tornadoes and the ground swell caused by seaquakes In the middle of the sixties, however, too many breeders overlooked an even more dangerous adversary—over-acquisitive haste. Prematurely harvested pearls lose their luster after only a few months. Over-production accompanied by declining quality temporarily landed the Japanese cultured pearl industry in a crisis.

127

Peanut pyramids on the outskirts of Kano in Nigeria. These pyramids — jute sacks tightly packed with groundnuts and piled up in genuine pyramid-building style without the aid of cranes or hoists — are found everywhere in the north of Nigeria, where more than a million small farmers harvest an average of 700,000 tons of peanuts per annum. But it is only in Kano, a sort of port for the waterless ocean of the Sahara, that they form such a large array. Kano also serves as a transit point for the peanuts harvested in Niger and processed overseas. The pyramids are consequently only temporary. They embellish the skyline of the old harbor town during the harvest, when the local oil mills and the narrow-gage railway connecting Kano and Lagos cannot cope with the incoming flow of nuts. Each pyramid (resting on a concrete base) contains more than 10,000 sacks of nuts weighing 600 tons in all.

128
Tobacco growing near Medan in Sumatra, Indonesia. The tobacco growers around Medan supply cigar manufacturers throughout the world with the best light wrapper leaf. Only one-seventh of the total area is under cultivation at one time; for the next six years the soil lies fallow to recuperate. In barns made of teak and palm fronds, the lower leaves are dried on bamboo frames. Apart from its general quality, its aroma and its even burning, it is the elasticity of the extremely thin Sumatra leaf that recommends it for the wrapping of cigars. The tobacco growers profit from advantages which their product owes to climate and soil, but they are now worried by their shrinking share of a shrinking market.

103

104

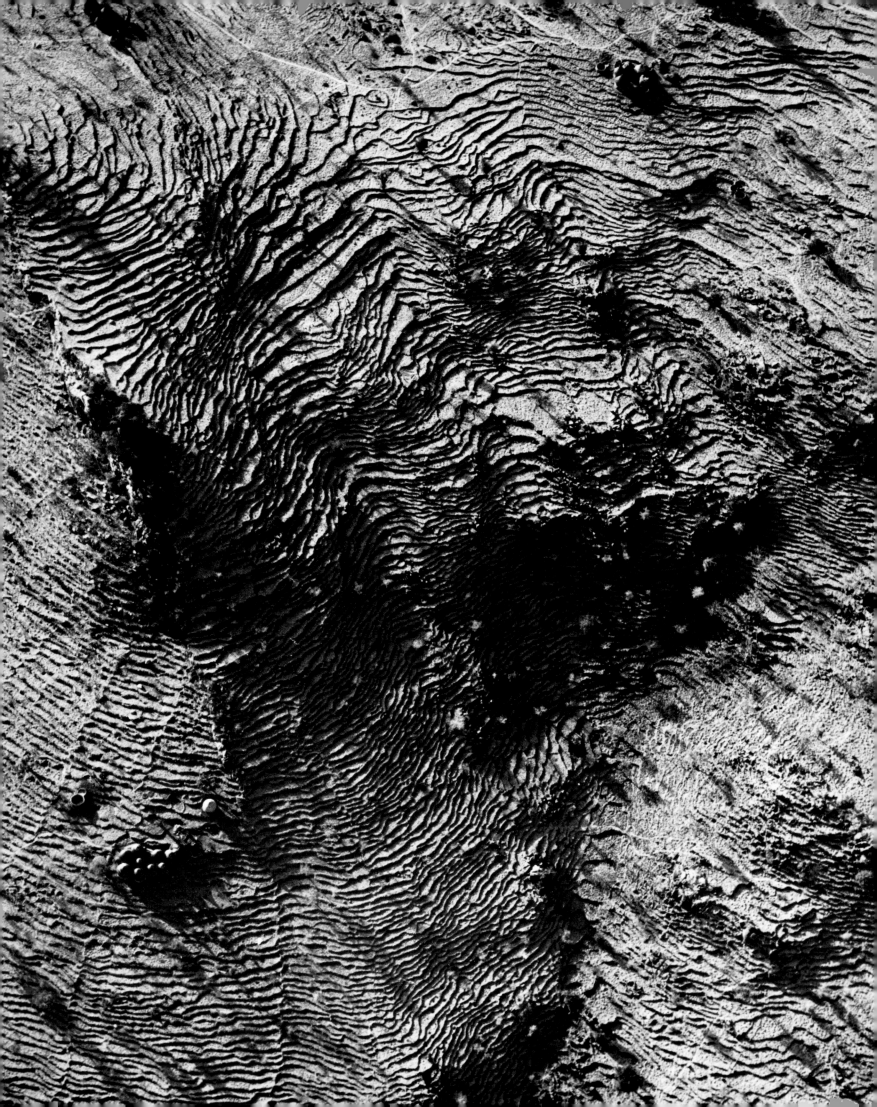

108

109
110 ▶

113

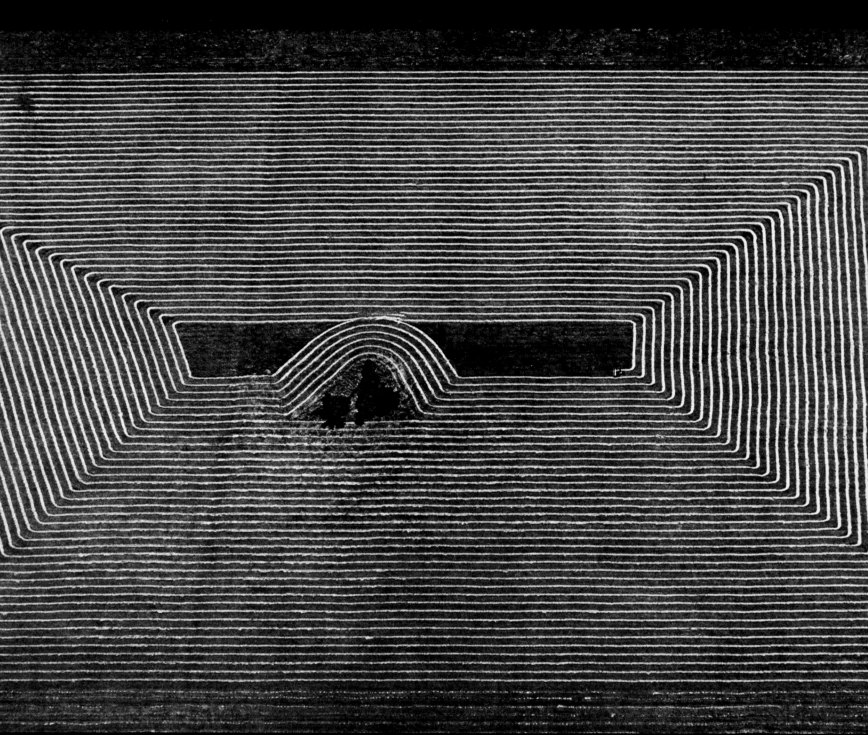

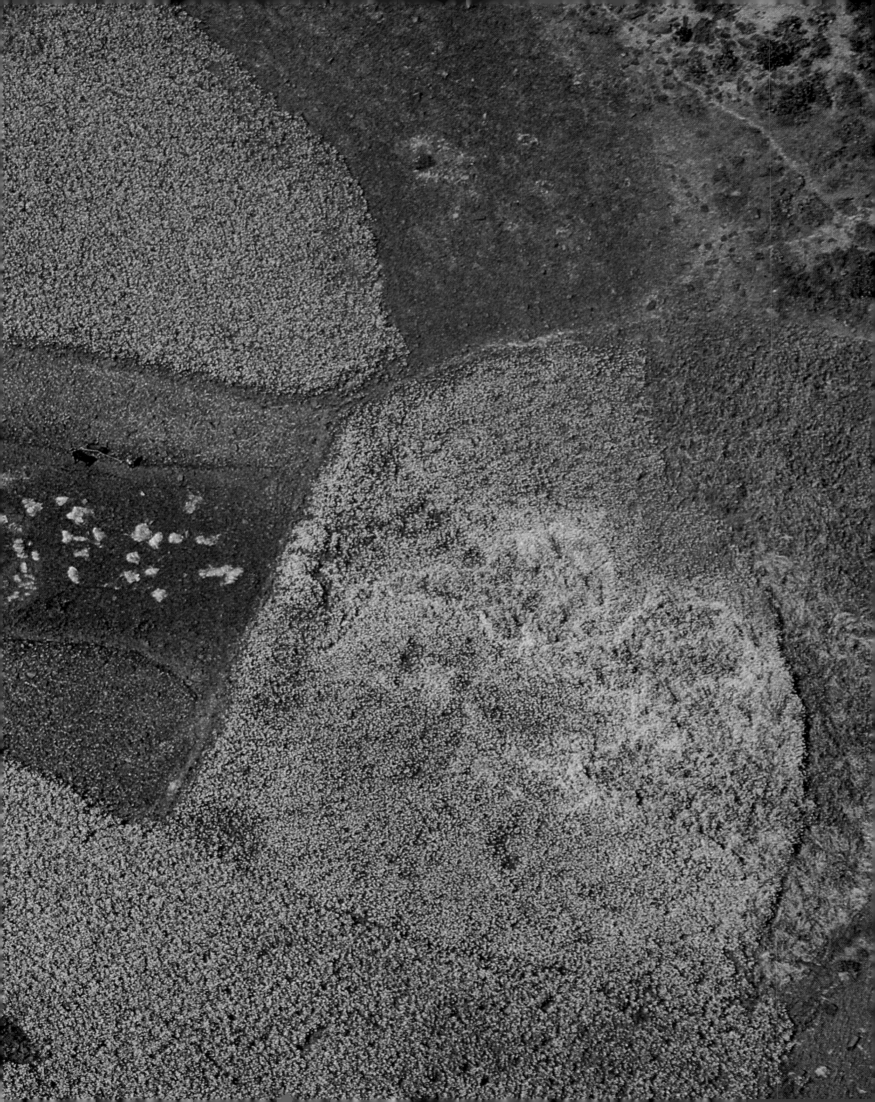

122 123
124▸

Flight over Nebraska

The Midwest, granary of America (and other parts of the world), appears to the traveler on the ground as a monotonous expanse broken only by church and water towers, grain elevators and silos. Dull? Certainly not from the air. The view from above transforms this landscape into one of the most visually exciting in the USA. And makes it clear that farmland is nature shaped by the hand of Man, that the farmer can — by purely technical production methods designed to maintain or increase his yield — change the graphic face of the landscape again and again in the course of thirty or forty years.

The optical delights of Nebraska derive from two sources. The first (and older) consists in the measures taken by American farmers, warned by bitter experience, to prevent the erosion and blowing away of the topsoil. The second (more recent) has resulted from the development of irrigation techniques enabling farmers to make better use of the fertility of the soil.

The most important of the conservational techniques is called strip cropping. On terraces, along the contour lines of cultivated areas, or at right angles to the direction of the prevailing wind, strip cropping prevents the loss of humus, stores humidity and increases harvests. In areas of low precipitation, where dry farming is practiced, the zebra landscape has emerged naturally — partly because fields have to be left fallow in summer to restore the moisture balance, partly because of America's right-angled surveying: the individual sections could only be divided into strips. In other parts of the country federal and state authorities have successfully encouraged strip cropping since the devastating dust storms of the 1930's. This practical concern for the soil has thus introduced visual tensions and beauties into the former monotony.

The farmer, then, as artist or designer? Does he appreciate the changing garb of his fields? He certainly has the means of admiring his creations from above, for many farmers use their own planes to go to work in the fields or shopping in town. The fact that here and there real strip orgies are celebrated makes one think twice about it. One can check all the factors that affect the original decision—for instance, direction of the wind, availability of mechanical equipment, steepness of slopes, run of contour lines, requirements of a particular crop, etc.—but the calculation never works out exactly. There is always a remainder, and that might well be creative irrationality.

The strips of the forties and fifties have now been joined by the irrigation plants of the sixties and seventies. They are characteristic patches on the face of America's new agricultural landscape. There can be no doubt about the success of these circular irrigation roundabouts in the arid regions of the U.S.A. Every flight from coast to coast reveals new irrigation works in the states of the Midwest and, beyond the Rocky Mountains, in California and the states of the Pacific Northwest. This irrigation system—*center-pivot irrigation* in the terminology of the specialists—was invented more than twenty years ago. With Patent No. 2604359 the American Patent Office protected Frank Zybach's "Self-propelled Sprinkling Irrigation Apparatus" on July 22, 1952. The inventor and his first licensees had to wait a long time for the success of their product, but for a good five years now the number of those who consider July 22, 1952, a day of revolution has been growing rapidly. At the present time some 12,000 plants are operating in the U.S.A. and two dozen manufacturers are looking for more clients.

The jet perspective reduces these irrigation turntables to the size of toys. In reality they are of considerable proportions, each an automatic farm in itself. The standard model can irrigate some 160 acres and larger models irrigate quite a bit more. The biggest— near Yuma, Colorado—covers a circular area of 520 acres.

204

The swivel arm with its spray nozzles rotating about the center consists of a number of sections. Each separate section is individually driven. If it moves too fast or too slow, a control signal switches the motor on or off and brings it back into place. In the majority of installations the discharge pipe of the swivel arm is supplied with groundwater from a borehole near the center. Some plants also utilize water from rivers and lakes. The sections of the rotating arm have hinged joints so that it adapts to uneven terrain and is not stopped even by terraces. The technically most advanced systems run on tractor tires (causing minimal damage to the soil) and are electrically driven. The farmer selects the desired speed from the control desk at the center of the roundabout, transmission being infinitely variable from 20 to 200 hours per revolution. He also sets the water supply at the well necessary for the desired density of irrigation, and the rest is up to the machine. It can even be programed to sprinkle only a particular sector of the circle and then switch itself off or turn around—for instance when the farmer is growing several crops with different moisture requirements in one irrigation circle, or when the circular irrigation area includes a large obstacle.

The advantages of these irrigation plants are obvious: they are fully automatic and require only one-tenth the labor of conventional irrigation systems. One man can operate up to thirty center-pivot irrigation systems. Irrigation specialists also stress the economical water consumption.

As could be expected, the costs of purchase and installation of these systems are high. In Nebraska an apparatus of standard size, together with a well 100 to 200 feet deep, sets a farmer back approximately $30,000 to $35,000. In the Midwest stupendous bankruptcies have occasionally occurred when farmers have failed to adapt their irrigation systems to the conditions of the soil—a danger inherent in all irrigation systems. Such inexperience resulted in catastrophic compression of the soil. Today the lesson has been learned, and irrigation systems that were originally reserved exclusively for valuable crops can today be economically viable even when the irrigated area is only pasture. Most banks now lend money on these plants without mortgage securities.

One farmer in the northeastern corner of Nebraska operates no fewer than a hundred irrigation systems. There are good reasons for the concentration of the plants in this state — Nebraska in fact owns one in three of the American total. The fact that the inventor himself lives there hardly counts for much. The real reason is the abundance of groundwater in Nebraska and the liberality inspired by this wealth. At present, drilling for water in Nebraska requires registration only and not an official permit.

129

Two center-pivot irrigation systems between North Platte and Imperial in Nebraska, one of America's most important farming states. In the dark halves of the circles alfalfa is growing, in the light ones corn. The farmer can select any combination of crops; his choice is limited only by the nature of the soil. Different moisture requirements of crops in the same circle are no problem with a center-pivot irrigation system. The swivel arm can be programed, for instance, to turn back after half a revolution.

130/131

Fields lying fallow near North Platte. In order to prevent the growth of weeds, improve the moisture content of the soil and counteract wind erosion, the uncultivated field is repeatedly worked during the vegetation period. The regular texture in the light areas has been left by a grubber. The farmer is now working with a tandem disk harrow, always driving across the slope of the terrain. The decorative border of the stubble field is produced by the turning tractor. Wheat was harvested here last season.

132

The farm of William Sturtevant, near Wauneta. It comprises 259 hectares (640 acres) of Sturtevant's own land, plus half as much again that he has on lease. Terrace cultivation is practiced to save water. The crops are rotated: wheat, corn and fallow, in that order. This photograph was taken in September. Wheat had already been sown on the summer fallow (brownish shades); on the dull green fields corn is ripening; and the harvesting of sorghum — brighter green and newly reaped fields — is in progress. To the north of the farm (left margin of the photograph) the Badlands begin; their runoff flows into a creek called Stinking Water, which flows into Frenchman River, a tributary of Republican River. Where the Badlands have a grass cover, they are used for pasture. Sorghum and some oats supply the winter fodder for the cattle, which also graze the corn stubble.

133–136

Works in an art exhibition — or "only" field graphics? For every line the farmer draws, for every pattern he creates, he has a good reason. Sometimes, admittedly, it is only an excuse. While the farmer was working with a disk harrow on a wheatfield lying fallow for the summer, he was forced by rain to interrupt his work for several days (Plate 135).

137

Growing wheat along contour lines, known as "contour farming." The dark fields are under wheat; between them are fields of wheat stubble which are lying fallow for the summer. They show traces of having been worked with duckfoot tines, which separate the roots from the stubble but leave the latter standing to protect the fields from wind erosion.

138

The agricultural landscape near Scottsbluff, early June: strip and terrace cropping of wheat and alfalfa. The light strips are lying fallow or planted with sugar beet which has just sprouted. The strips at right angles to the direction of the prevailing northwest wind. The direction and size of the strips are also influenced, however, by the terrain, the condition of the soil and the farmer's mechanical equipment. The two irrigation circles, where corn is being grown, are new and the swivel arms and sprinkler nozzles have not yet been mounted. Each square enclosed by the roads is a "section," one square mile in area.

139

A farm near Imperial. It consists of a total of fifteen center-pivot irrigation systems, each of which occupies one-quarter of a section. (The same farmer owns two further farms of a similar size.) Corn is growing on the light circles, grass and alfalfa on the dark ones. On two the alfalfa is just being harvested, while in a further circle—a segment of which is visible at the bottom of the picture—the farmer is growing potatoes.

140

A center-pivot irrigation system growing alfalfa. Exactly controlled irrigation permits up to four harvests per year. At full speed (one revolution in twenty hours) and with a water supply of 5,000 gallons per minute, a standard ten-limb system supplies about 0.17 inches of water to an area of 135 acres. At very slow speed (one revolution in two hundred hours), the amount of water can be increased almost tenfold.

The systems run above the ground and can be used even with tall plants—in fact, corn is grown under every second irrigator in Nebraska. The supports of the swivel arm describe concentric circles in the field. The well and pumping installations are outside the photograph, a short distance from the center of the system. This new irrigation technique has also been a success in other countries. Among the users are Mexico, Australia and Libya; the last-named is using the technique to exploit extensive groundwater supplies near the oases of Kufra, in the midst of the desert.

141

A bare fallow field—previously planted with corn or sugar beet—has been worked over with a rotary hoe to prevent the blowing away of the topsoil. Evidently this measure has been only partially successful, for the aerial picture shows traces of drifting soil here and there. The farmer hopes for rain. If he is disappointed, he will have to have recourse to other agricultural machines from his defense armory to deepen the furrows or raise their edges.

129

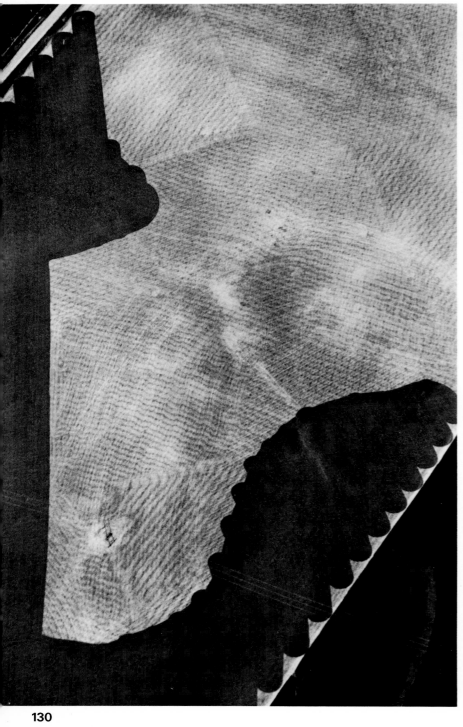

130

131

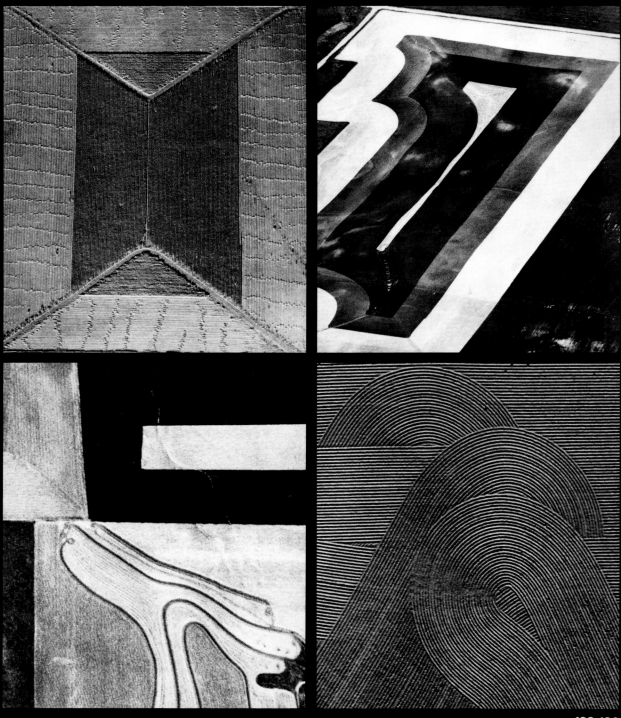

133-136

139

140

Archaeologists take to the air

The first aerial photographs ever published of an archaeological site were taken by a certain Lieutenant P. H. Sharpe. During a practice flight in 1906 he photographed from his war balloon the most venerable and mysterious stone circle left behind by Europe's prehistory—Stonehenge (Plate 168). His photographs caused quite a stir. Sharpe, however, was not a farseeing innovator: Stonehenge had only happened to be included on one of his plates.

The first aerial pictures of an archaeological excavation were taken by Sir Henry Wellcome, an amateur archaeologist. Shortly after 1910, Sir Henry, a passionate handyman, launched a box kite into the air above his private excavations in the Sudan—not forgetting to attach a remote-controlled camera.

The very first time that archaeologists systematically used airplanes to photograph places of archaeological interest was during World War I. This was also the time when the airplane first assumed its (now preeminent) role in limes research, i.e., exploration of the border fortifications of the old Roman provinces. After 1916, Theodor Wiegand, Inspector-General of the Antiquities of Syria, Palestine and Western Arabia, had pilots of the German Air Force Unit 300 take photographs of ancient towns for the German—Turkish Detachment for the Protection of Monuments. At about the same time Carl Schuchhardt persuaded German military airmen to take aerial photographs of the so-called Trajan Ramparts in the Dobruja (Rumania). Friedrich Sarre satisfied his ambition to obtain aerial photographs of Mesopotamian ruins, namely ancient Samarra (Plates 147 and 169), a venture he had unsuccessfully attempted before the war with the aid of a kite. These German pioneers deserve honorable recognition; but the promise of archaeology from above was already literally "in the air."

219

On the side of the Entente there were likewise many pilots who were interested more in the location and recording of old ruins than in the creation of new ones. The English Lieutenant G. A. Beazeley could hardly contain his astonishment at the irrigation systems and remnants of settlements that were revealed to him during flights over Mesopotamia. Before he was shot down and captured he was thus the second to "discover" Samarra. The French Jesuit priest Antoine Poidebard, a navigator-observer during the war, was also sufficiently stimulated to become actively involved later in fruitful aero-archaeological activities in the Near East and in North Africa.

For the archaeologist the plane is a unique means of getting an overall view. It is, so to speak, an elevated strategic position where nature may not offer him one. At the outset, the plane's role in archaeology was limited to supplying a first quick overview and replacing laborious ground surveys, thus saving both time and money. Soon it became obvious, however, that aerial photography could also be used in the search for archaeological sites, revealing surface marks that are not visible or not amenable to interpretation on the ground. The aerial picture turns surface irregularities into meaningful patterns. It even lends the archaeologist x-ray eyes with which he can see *beneath* the earth. He can make out the ground plans of ancient towns, houses, fortifications and tombs; follow procession paths and roads, old systems of field cultivation and divisions of land long buried by the plow. These traces show up when oblique light plastically models even the slightest unevennesses of the surface, or when discolorations of soil or vegetation betray the things that once have been. The Englishman O. G. S. Crawford, a reconnaissance airman over the western front during World War I, but too preoccupied with survival to go in for archaeological studies, founded in the 1920's a branch of archaeology that uses the aerial picture mainly as a means of locating possible archaeological sites. Together with Alexander Keiller, he published in 1928 a book entitled *Wessex from the Air*. The birthplace of this new method was not accidental. In the part of southern England that Britons still like to call Wessex after the old Anglo-Saxon kingdom, and particularly at its heart, in the counties of Dorset and Wiltshire, there is an overwhelming abundance of prehistoric remains, many already faded on the ground but as clearly visible from the air as on the day of their creation. The scarcity of forests in southern England facilitated archaeological research from the air. And here again, war was responsible, at least indirectly, for Crawford's and Keiller's epoch-making photographs: the two British pioneers took them with a camera captured from the Germans.

Under Crawford's guidance archaeology took off on its flight into the past. He was the first to speak of *shadow sites*, revealed when slanting light casts shadows of slight elevations; of *crop sites*, which give themselves away by a visible difference in vegetation, for instance where poor wheat growth marks the position of buried walls; and of *soil marks*, discolorations of the soil, which are mostly due to moisture differences caused by underground ruins. Crawford's categories, though technically refined and expanded upon, are still valid today.

It does not detract from Crawford's achievement to point out that he was not, in fact, the first to notice such things. Circular patches of sparse vegetation have long been noted above underground burial mounds in northern France, for example. For centuries these anomalies were explained by the inhabitants as *danses de fées*, dancing places for nocturnal fairies or goblins. But by the seventeenth century the Carinthian Johann Dominikus Prunner of Sonnenfeld was able to interpret correctly the uneven growth of buckwheat above the walls of the Roman town of Claudium Virunum (near today's Klagenfurt in Austria). In the absence of a

221

flying machine he had to be content with the view from a local mountaintop. The same talent for observation and the same intelligence, again exercised by looking down from a mountain or hilltop, was later demonstrated by English and French amateur students of antiquity — ingenious representatives of aerial archaeology *avant la lettre*. The American balloonist John Wise, on his 132nd balloon flight over Chillicothe, Ohio, in 1852, clearly perceived the contours of earthworks, but later, when he went to look for them on foot, could find nothing. He concluded that certain discolorations of the soil could be distinguished only from balloon heights, and he drew attention to the advantages of observing the earth from above. Finally, Sarre too realized the significance of ground signs when he analyzed the aerial photographs of Samarra. The streets and buildings of the town were visible even where the mud walls had been demolished. Traces of walls that had escaped topographers were revealed by "the discoloration of the soil and the differences of the vegetation."

142

Barrows dating back to the Bronze Age in a field near Avebury, Wiltshire, England. The traces left by harvesting machines reveal the respect—and despair—of the farmer. Hundreds of such mounds have been plowed over and survive at best in the form of crop and soil marks on aerial photographs. The arch-enemy of English archaeologists is not the bulldozer but the tractor and plow, which break up ground previously used for pasturage. The integration of England with Europe threatens on an alarming scale the incredible range and variety of the British prehistoric heritage.

143

The hill fort of Hod Hill in Dorset, England. Whenever the sheep-nibbled turf on hills in the south of England is thrown into relief by a rising or setting sun, Iron Age fortresses are dramatically revealed as an integral part of this landscape. The archaeological map of the southern half of England shows nearly 1,400 such hill forts, which, if not fortresses in the modern sense, were at least fortified enclosures, mainly dating from the second part of the Iron Age (350 to 150 BC). The immense amount of timber used for framework and palisades in the ramparts of these fortresses contributed considerably to the radical deforestation of the English South Downs. The Romans occupied many of these forts after the conquest of Britain, turning places of refuge into the citadels of dominion. In Hod Hill, too, the Romans built a *castrum* for cavalry and infantry in the northwestern corner of the fort in AD 45. From here the new overlords supervised the original builders of the Celtic fort, the Durotriges.

144

Knowlton Rings in Dorset, England. The double ring of rampart and moat was constructed in 1800 BC. The fact that the moat lies inside the rampart (as with most of the other henges from the Bronze Age, but not with the most famous of them, the one from which they take their name—Stonehenge, Plate 168) makes it clear that the fort was not meant for defense purposes. The enclosures are inspired rather by cultic and magic motives. The twelfth-century Norman church—today in ruins—at the center of the circle broke the spell. Or might it be that it sought safety within its bounds? Even in the ground plan of Canterbury Cathedral, or rather of its forerunner, destroyed by the Normans, researchers claim to recognize megalithic measurements and the vestiges of ancient stone circles.

145/146

Carnuntum, a Roman fort on the Danube frontier, between today's Deutsch-Altenburg and Petronell, Austria. Emperor Marcus Aurelius (161—180) fought the rebellious Marcomanni from Carnuntum, and wrote here the second chapter of his *Meditations*. The pictures show ground plans of streets and buildings in the camp village lying in front of the legion's camp proper and divided from it by the *glacis* (the combat area). The village grew under civil administration into a town in its own right. The remains of underground walls, which cannot be detected at ground level, show up in the aerial photograph as dark lines in the ripe wheat; but they completely disappear where the crops are still green. Only wheat shows this very sensitive reaction to local variations in the moisture of the soil: the lime and mortar of the Roman walls extract moisture from the soil above them so that the plants grow less high and ripen earlier. The resolving power of this archaeological radiography is good enough to reveal underground walls only one foot thick.

147

The palace of Caliph al-Mutawakkil (847–861) in Samarra, Iraq. The immense ruins of this great city (which once boasted a population of over a million) extend over twelve miles along the eastern bank of the Tigris, but they are now visible only from the air. In their search for cheap building materials, the local inhabitants have carried off the old walls except for a few remnants (see Plate 169). The history of ancient Samarra was incredibly short. The Abbasid Caliph al-Mutasim, son of Harun al-Rashid, made it his new residence in 838, being tired of Baghdad. An extravagance that was unique at the time catapulted the town to fame. The palaces of al-Mutawakkil, in particular, caused a great stir. The picture shows his last palace, at the northwestern end of the town. Inside a pentagonal wall enclosing 250 acres of ground were throne rooms and private apartments, baths and pavilions, gardens with fountains, buildings for harem and courtiers, accommodation for guards and cavalry. . . . On the right of the picture is one of the two canals that enclosed Samarra and its suburbs, turning the city into an island. Long straight rows of ventilation and maintenance shafts reveal the course of the underground tunnel for the water supply, the *qanat* (Plate 111). Samarra died with al-Mutawakkil, for his successor transferred his residence back to Baghdad. It had moved like a meteor across the sky of the early Islamic world. Many contemporaries saw its decline as a punishment from heaven: al-Mutawakkil had wickedly destroyed the tomb of Husain, grandson of the prophet, in Karbala.

148

Sigiri ("Lion's rock") of King Kasyapa (fifth century) in Sri Lanka (Ceylon). As a rock landmark his residence, 590 feet above the emerald-green jungle, has no equal. King Kasyapa's architects not only used the mountain as a base for their buildings but converted it into an architectural sculpture. Even the huge fragments of rock lying immovably at the foot of the bastion were included in the general design. A colossal lion of bricks and plaster—the remains of its paws can be seen on the photograph, lower center—guarded the entrance to the palace on the summit. King Kasyapa seems to have had in mind a kind of celestial mountain, perhaps as a pedestal for his own majesty. He had the whole western side of the mountain plastered and may have had it painted with pictures of nymphs. Kasyapa was a usurper and a parricide. Yet Sigiri is no gloomy place of refuge but a palace of love and delights—with swimming-pools, pleasure-houses, baths and pavilions. When the avenger, a half-brother of the king, approached from India, Kasyapa left his retreat for a trial of strength on the battlefield. There his luck deserted him, and in order not to fall into his brother's hands, he killed himself.

149

Masada ("rock stronghold"), 2.5 miles from the shore of the Dead Sea and 1,500 feet above it, in Israel. Herod turned the castle on the eastern slope of the mountains of Judaea into a huge fortress. He had a casemate wall erected around the summit, built towers, storehouses, barracks, cisterns, armories and luxurious accommodation. The photograph shows the remains of the royal storehouses and baths on the castle rock and the three-story "hanging" palace of the king. It is no accident that Masada is today a symbol of the will of the modern Jewish state to defend itself. In the Jewish war of liberation against Rome (66–73), Masada was the last stronghold. When its position became hopeless, in 73, the besieged — 960 men, women and children — took their own lives so as not to be captured by the Romans. The Bible does not mention Masada.

150/151

Snapshots of the race against the organized deluge caused by the construction of the high dam near Aswan. Egypt was prepared, at the worst, to drown the past to save the future. Before doing so, however, it appealed to the cultural solidarity of nations. In the sequel, the Nubian Nile experienced the most expensive move of all times, with international participation: people and monuments were given new homes away from the storage lake or above its highest water level. At the same time archaeologists researched five thousand years of remains left by half a dozen cultures and civilizations. Plate 150 shows Jebel Adda, a Meroitic town, which remained important even in Nubia's Christian period up to about 1500. When the picture was taken it was already the eleventh hour for the rescue and excavation work. The artificial lake had separated the acropolis from the mainland and was now licking at the Meroitic, Christian, Fatimid and Mameluke necropoles lying at the foot of the acropolis. Plate 151, also dating from the spring of 1968, shows the re-erection of the great cave temple of Abu Simbel in a technically tricky phase. The temple, which had been sawn out of its mountain location in individual blocks, was being reassembled inland at a higher altitude — out of danger of flooding. The colossal seated figures of the god-king Rameses II once more smiled from the façade upon the ever-widening Nile. The engineers were erecting the concrete cupola structure consisting of a cylindrical collar at the front and a spherical rear part, which was to distribute the weight of the artificial hill without endangering the temple.

152

Old field systems near the town of Turi in the Atacama Desert, Chile. Turi, situated 10,000 feet above sea level in the foothills of the Cordilleras, was the most important settlement of the Atacamenos in pre-Columbian times. Approximately a century before the Spanish conquest, the Incas overran the whole of North Chile. Traces of their dominion have remained in Turi to the present day. The skeletal pattern of extensive field systems outside the town, for instance, is no doubt due to these keen tillers of the soil who swept down from the highlands. The fields have since been covered by the drifting sand, but their outlines are still visible from above.

225

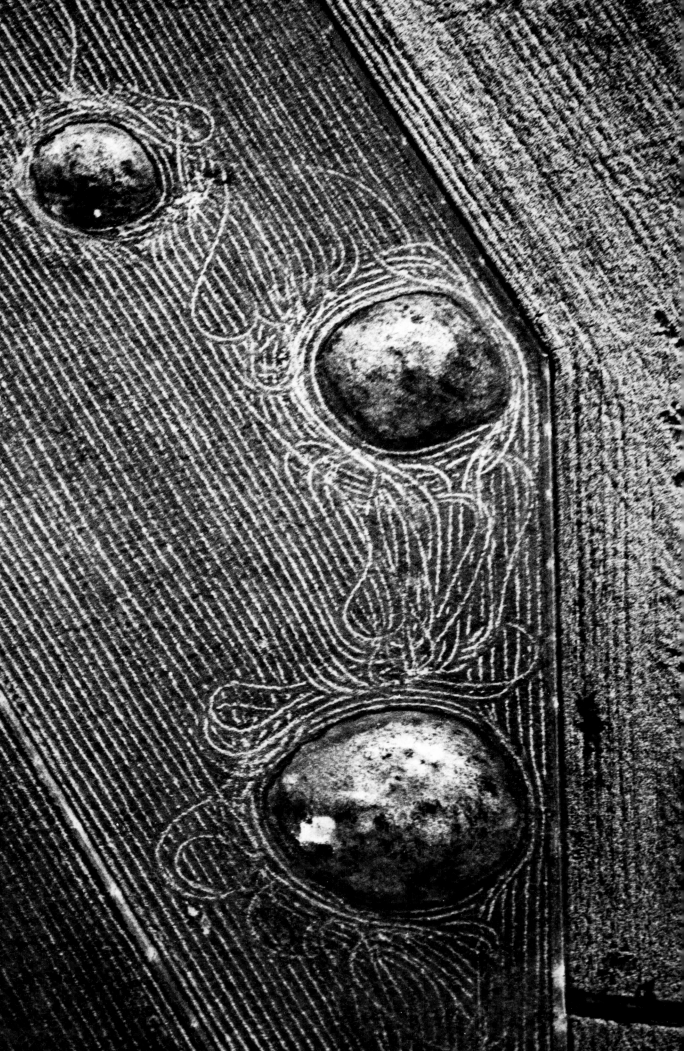

144

145 146
147 ▶

148

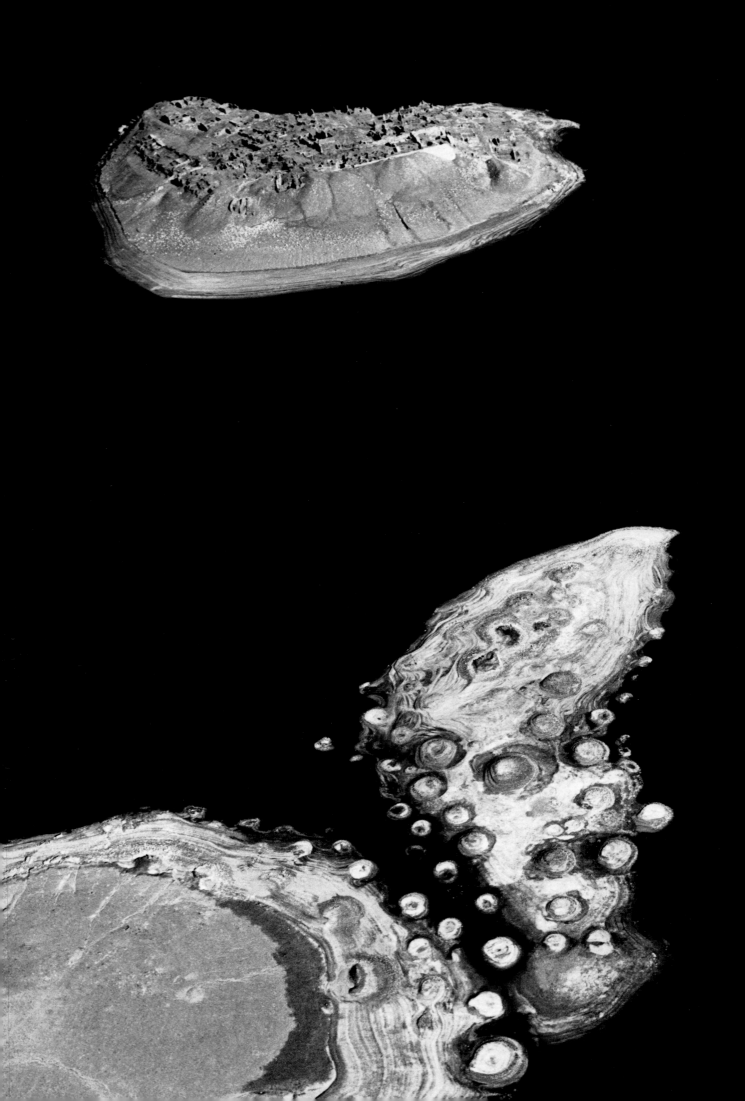

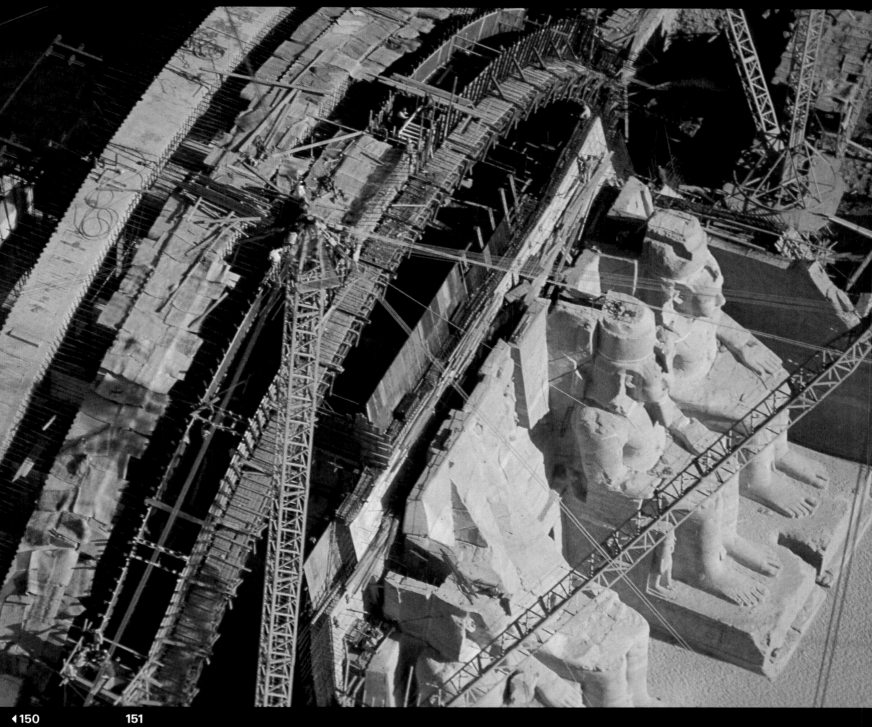

Biblical sites
and cities from the air

Palestine was among the first places to be photographed from the sky. During World War I airborne photographers of the German—Turkish Detachment for the Protection of Monuments operated over the Sinai region and the Negev. Since then, the airplane—or sometimes just a balloon or a kite—has become indispensable for archaeological work, particularly in a country where high places have always had a religious significance. Like archaeologists today, the kings of that time were aware of the advantages offered by an elevated standpoint. But how could we today, without aviation, look down on the buildings they left behind—at the acropolis of one of Solomon's chariot cities, or Herod's castle and tomb?

The flight into the past is more rewarding over Palestine than anywhere else. It opens up an incomparable source of historical knowledge, which is illuminated in turn by the Bible. A flight over Palestine is memorable, too, for the beauties of the landscape. The hills of Samaria shine like pure silver. Galilee appears as a carpet of flowers. In the distance the snow-covered Hermon towers over Mount Tabor: "Tabor and Hermon shall rejoice in thy name" (Psalm 89). And the mountains of Judaea (Plates 103 and 104) sing the praises of the farmer's toil.

239

153

The Jebel Musa ("Mountain of Moses"), a summit in the granite wilderness of South Sinai. Rising 7,500 feet above the level of the Red Sea, it is not the highest mountain of the area, but for sixteen centuries the Christian tradition has venerated it as the holy mountain upon which Jehovah appeared to Moses and made his covenant with the tribes of Israel. Snow fell on the day before the photograph was taken — a sprinkling only, but unusual even for a December day.

154

The ruin-strewn hill of ancient Jericho (Tell es-Sultan), West Jordan. The whole of the hill, which is sixty-five feet high, consists of the rubble of earlier civilizations. No other town on earth can compete with Jericho in age. In the deepest layer, opened up by the exploratory trenches visible on the photograph, were discovered the oldest stone houses and town fortifications known so far — dating back to the eighth millennium BC. Fundamentalists who claim that the Bible cannot be wrong are admittedly out of luck in the case of Jericho. Around the time of the occupation of Canaan by the Israelites (between the fifteenth and thirteenth century BC) Jericho, though already over six thousand years old, would seem from the excavations to have declined to insignificance. In the previous seven hundred years, Jericho's walls are known to have been destroyed and rebuilt — or at least mended — no fewer than twenty times. But of the walls which, according to the Bible, Joshua's trumpets brought tumbling down the archaeologists can find no trace.

155

Et-Tell near Deir Diwan, in the mountains of Judaea, West Jordan. The ruins are generally thought to be the remains of Ai which, according to the Biblical report, was the first town Joshua took in the mountain country, when he "made it an heap for ever." The excavation evidence contradicts this version. When God's chosen people settled in Canaan, the town had been lying in ruins for a thousand years. The contradiction might, of course, be due to the fact that the identification of et-Tell with Ai is erroneous.

156

The hill with the ruins of the Philistine town of Ekron, Israel. Spring flowers effectively set off the tell against the surrounding cultivated land; the rectangular walls of the acropolis shine through the flower carpet. According to the Book of Judges, Ekron fell after the death of Joshua; but it was conquered, if conquered at all, only temporarily. For a short time the Philistines even put up in Ekron the Ark of the Covenant, which they had captured. Later, after Goliath's death, they sought refuge in the safety of the town. Ekron only came permanently into the hands of the Israelites in the second century BC, when the Syrian king Alexander Balas presented it as a gift to Jonathan Maccabaeus. Modern excavations have yet to be undertaken in Ekron.

157

A town without its like: Jerusalem. Not the heavenly Jerusalem the soul seeks for, but the Jerusalem of history conquered, torn down and rebuilt, a bone of contention to this day, earthly and in the news. The rays of the setting sun over the old part of the town light up, on roofs already in the shade, the forest of television aerials which sprang up after the war of June 1967 and the reunion of the divided town. King David took the Jebusite fortress on the heights of the Judaean hills and from there ruled over north and south; Solomon made the town rich, Herod made it great. For nineteen centuries after the destruction of the second temple it remained a focal point of hope for dispersed Jewry: "Next year in Jerusalem!" The Christian churches, too, see more in Jerusalem than just a hotly contested town at a crossroads of history: here, where God's son died and rose from the dead, time and eternity meet and merge for the believer. And even this is not the end of its holiness: Mohammed, when in flight, sometimes told his disciples to turn their faces to Jerusalem instead of Mecca in prayer—and it was from the temple hill that the prophet rode up to heaven to receive illumination from Allah. The present-day walls were built by Sultan Suleiman the Magnificent. In the southeast (top right in the photograph) can be seen the Haram ash-sharif, the former temple square, where since the seventh century the Dome of the Rock stands (over the rock of Abraham's sacrifice and of Mohammed's heavenly vision) and since the eighth century the Aqsa Mosque. The evening sun falls on the Wailing Wall, a part of the western retaining wall of the Herodian temple, which has been freed of its debris since 1967 and is now accessible across a wide square. At bottom right in the picture is the complex of Christian sanctuaries, above all the place of Christ's crucifixion and burial (under the domes of the Church of the Holy Sepulchre). Of the ten measures of beauty that descended on the earth, Jerusalem claimed nine. But the Babylonian Talmud says that nine is also Jerusalem's share of the ten measures of sorrow. And politicians have done everything in their power to preserve the rightful proportions—if not those of spiritual beauty, then at least the town's share of its only too palpable sorrow.

158

The tell of the town of Megiddo at the western entrance to the plain of Jezreel, Israel. Excavations have exposed palace ruins, fortifications, a water supply system designed for times of war, stables and storehouses. Solomon fortified Megiddo—made it a district capital and garrison for chariot troops—so as to safeguard the Jordan crossings and control the caravan routes between the valley of the Nile and Mesopotamia.

159

The acropolis of the town of Hazor, Israel. The road from Tiberias to Metulla runs round the venerable tell on the eastern edge of the upper Galilean mountains. Hazor was the capital of the northern Canaanite territories; the Israelites conquered the town, and Solomon turned it into a fort. In the upper half of the picture are citadels, ruins of royal storehouses and, between them, the rectangular shaft that gave access to the drinking water.

160

The ruins of the town of Gezer, on the edge of the hill country between Jerusalem and the Mediterranean. A sap—reminder of recent hostilities—marks the northern and eastern edge of the hill. Above the center of the picture, on the left, the ruins of a monumental gate from the time of Solomon; and to the right of it, where archaeologists have filled the sap with excavation debris, a casemate of the inner fortifications of the town and a square with *mazzeboth* (stone pillars) which may commemorate a solemn alliance. Gezer, a royal residence of the Canaanites from the third millennium BC, controlled the routes between Egypt and Assyria. Solomon obtained it as part of the dowry of the daughter of Pharaoh whom he took into his harem. He fortified the town against Ekron (Plate 156). The famous calendar of Gezer is a limestone tablet found there, presumably dating back to Solomon's times and listing the twelve months with information about sowing and harvesting.

161
The escarpment of the mountains of Judaea looking toward the Jordan depression and the Dead Sea: the wilderness of Judah with the old road to Jericho. The gorges and caves of this region have offered shelter to the banished and persecuted, to preachers and robbers, pioneers and footpads throughout Palestine's tumultuous history. At the top left edge of the picture, behind the haze, the oasis of Jericho and the Dead Sea.

162
Samaria on its mountain, West Jordan. Samaria is one of the very few towns founded by the ancient Israelites. Omri, one of the most enterprising kings of the northern territories (though probably of Arab extraction), built Samaria as his capital around 870 BC. Excavations have revealed ruins of the Israelite king's palace on the mountain and town fortifications belonging to the same epoch. The forum, a basilica, a theater and a columned street — from which the archaeologists have also removed the debris of the millennia — recall the town's golden years under Herod and the Romans. Outside the left-hand edge of the picture lies the Arab village of Sebastiye. Its name perpetuates a piece of flattery for Herod, as under his rule Samaria was rechristened Sebaste from Sebastos, Greek for Augustus, Herod's imperial benefactor.

163
The hill country of Samaria, West Jordan. In the Old Testament this central region of the West Jordan highlands was known as the mount of Ephraim.

164
The Herodeion on the threshold of the wilderness of Judah, West Jordan. Herod maintained a palace and baths on the mountain during his lifetime, and it is probable that he intended the place to be his monument from the first. Between 22 and 15 BC he had the natural hill remodeled and banked up to form a regular truncated cone, probably wishing to imitate the mausoleum built a few years previously for the Emperor Augustus. In the Jewish war of liberation against the Romans the Herodian fortress was one of those that still held out against the besiegers when Titus and Vespasian, after taking Jerusalem and destroying the temple, were celebrating their triumph in Rome.

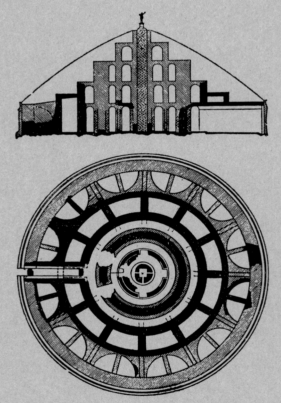

The mausoleum of Augustus (after *Segal*).

153

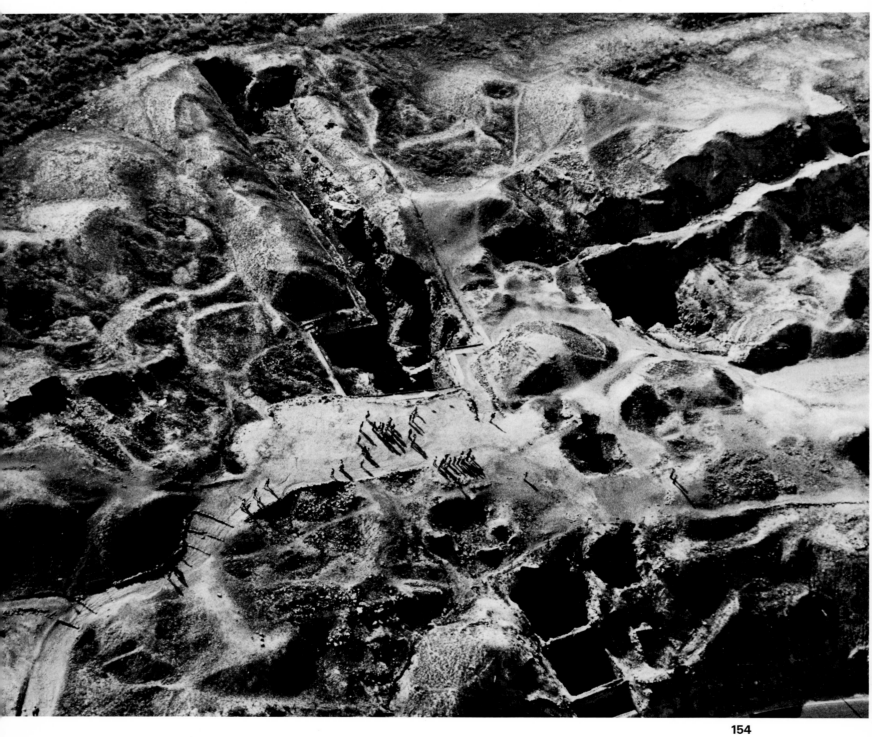

154

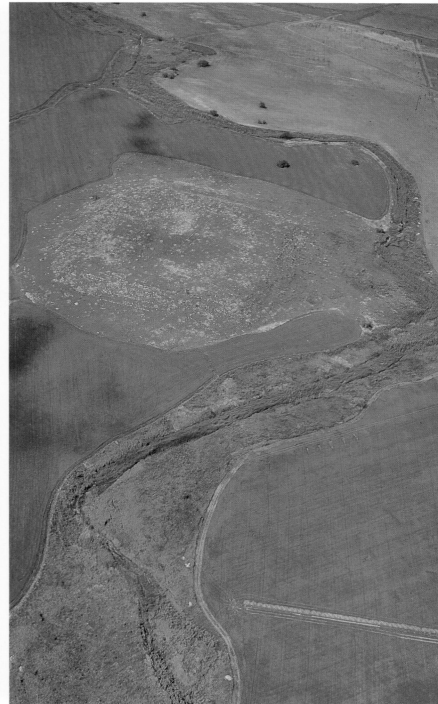

155

156

157 ▶

158

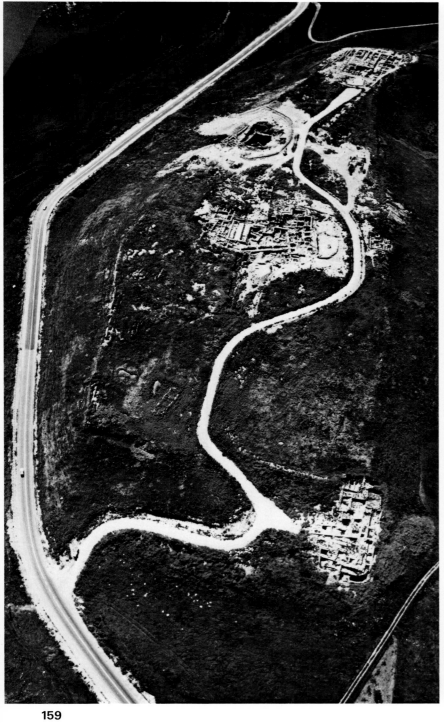

159

160

161 ▶

162

163

164

Monumental question marks

How far from the earth can an astronaut travel without losing sight of all signs of earthly life? And how near must a visiting astronaut come to be able to recognize signs of life on our planet?

The earth's signals, like those of a radio star, penetrate a few hundred light-years into space, and could be received by extra-terrestrial intelligences even if they are no further advanced in communications technology than modern earth dwellers. For the last few decades at least, the irregularities of the earth's radio signals must have revealed to possible neighbors in the Milky Way that this is an inhabited planet.

And the reflected light? To the astronauts on their way to the moon, still in the terrestrial field of gravity, the earth appeared like a marble on the black velvet of space — colorful but cold, with clouds, oceans and land masses suggesting the possibility of life but supplying no evidence whatever. Carl Sagan, head of Cornell University's Laboratory for Planetary Studies and spokesman of the American exobiologists, has studied thousands of photographs obtained by meteorological satellites for indications of life. He summed up his investigation in a challenging paper that asked the question: Is there life on earth? From space, said Sagan, it is impossible to perceive separately two points lying only 100 yards apart; therefore, the traces left by man's activities cannot be satisfactorily distinguished from geological features. Only when resolution is better than 100 yards can such traces be recognized for what they are. But even then there may be misunderstandings about the type of life assumed to exist on our planet. Sagan once said, only partly in jest, that Martians might well take the automobile to be the principal form of life on earth. After all, the environment is changed to meet its requirements; it moves, eats and ejects the products of its metabolism; it is looked after by an army of two-legged slaves, among whom it regularly selects victims for ritual slaughter. . . . We have to take a step nearer to the earth, and then suddenly the patterns of living creatures are joined by their monumental question marks.

165

Ayers Rock in Central Australia. The rock haunts popular tourist literature as "the biggest monolith on earth." Not only is it not the biggest, but it is not a monolith at all: rather an inselberg of stratified rock (arkose). For local tribes of the Australian aborigines, Ayers Rock is a mount of the gods, totally enveloped in legends and myths. Caves, shelters and other minor features of the rock are associated with memories of cultural heroes of the "dreaming" time, and sacred rites continually make this time operative in the present. The first and last rays of the sun set the mountain on fire—one can hardly put it otherwise. Its spectacular color changes have made this rock in the red, dead heart of the continent into more than just an ordinary destination for sight-seers, even for white Australians, and an excursion there takes on the dimension of a pilgrimage.

166

The Church of St. George (Beit Giorgis) in Lalibela, Ethiopia. Lalibela's monolithic churches rank high in the catalogue of the wonders of the world. Cut from a single rock, yet reproducing architectural forms in their internal and external design, they are in fact huge sculptures—architecture by subtraction. They stand at the bottom of pits, trenches having been dug around them to allow work to be done on the living rock. It is assumed that Beit Giorgis, the last of Lalibela's monoliths, was built at the beginning of the thirteenth century. Its elegance is unmatched. The cruciform building, with three crosses inside each other as a roof decoration, stands on a three-stepped cross-shaped base. The church above this base was carved from a rock thirty-five feet high and forty-one feet square. Passages lead down to the bottom of the excavation. The church

interior is a cruciform space undivided by pillars.

167

Mont-Saint-Michel in the Département of Manche, France. Low tide has exposed the sandbanks around the granite island. This innermost niche of the Gulf of Saint-Malo has continental Europe's largest tidal amplitude (up to forty-six feet); at neap tide the sea uncovers up to nine miles of sandy flats. Despite its uneven course, the tidal inlet at the foot of the hill was for centuries the boundary between Brittany and Normandy. Water engineering measures at the beginning of this century, however, robbed the Bretons of "Le Mont," to their persisting mortification. The presence of monks' settlements on the granite crest can be traced back as far as the eighth century, and its Benedictine history begins in the tenth. The buildings with the monastery church that crowns them were erected between the eleventh and sixteenth centuries. No besieger has ever conquered the abbey. Nevertheless, this prayer in architecture has known times of humiliation. Temporarily it served as a prison; later it became a museum. Since 1966, three monks have been living on Mont-Saint-Michel again. They assemble speedometers for a nearby factory.

168

The sacred stone circle of Stonehenge in Wiltshire, England. An African statesman not long ago wished (rhetorically) to take Stonehenge home with him as a proof that even Europe had gone through primitive stages of culture. However undisputed this statement is, Stonehenge is far from being a good illustration of it. Its concentric circles with their two inscribed horseshoes were created by what was then—

257

the first half of the second millenium BC—
a monumental effort. Its opening toward
the point of sunrise at the time of the
summer solstice, together with the alignment of the stones according to certain
solar or lunar phenomena, showed Stonehenge to be a center of sun worship. In
the 1960's, a Harvard professor named
Gerald S. Hawkins offered a new solution
to this petrified conundrum from European
prehistory. A modern high-speed computer helped the astronomer in his sur-
prising new assessment. The stone monument, said Hawkins, is itself a sort of New
Stone Age computer, with the aid of which
astronomer-priests were able to predict
exactly the succession of the seasons and

169

The spiral minaret of the Great Mosque of
Samarra, Iraq. The ramp spirals up in five
coils from the base to a cylindrical top
story. The base measures 108 feet square,
and the topmost platform is 164 feet above
the base. During the siesta natives come
here for cooling shade. Caliph al-Mutta-
wakkil (847—861), who was unrivaled as
a builder in Samarra (Plate 147), took up
the Sumerian-Assyrian-Babylonian tradition of the sacred tower. The spiral minaret
is accordingly an early Islamic echo of the
Bible's criticism of the presumption of the
Tower of Babel. With it the Caliph boldly
countered the misunderstanding caused
by the Biblical polemic and to a similar
allusion to Babylonian towers in the
Koran. The ziggurats were never fists
rebelliously shaken against heaven; instead, the temple towers were an invita-
tion to God to descend and live among
needy mortals. The circular shape of the
Malwiya ("the spiral") stresses the desire
for cosmic unity.

170

The Tower of Babylon, Iraq, on the site of
the ruined city. This picture shows the results of plundering of the coveted fired
bricks by local inhabitants in the nineteenth century. Groundwater instead of
burnt bricks now encloses the central part
of the ziggurat, which is made of unburnt
clay bricks. The tower is today disrespect-
fully called "the saucepan"; the handle is
formed by the trench of the old stairway.
The date of the founding of the ziggurat of
Babylon—called Etemenanki ("foundation
stone of heaven and earth")—is unknown,
as its lowest parts lie in the groundwater
and cannot be excavated. The oldest
archaeological finds date from the reign
of Esarhaddon (680—669 BC); the flores-
cence of the sanctuary began with Nebu-
chadnezzar II (604—562 BC), the sacker
of Jerusalem. In this New Babylonian
version of the ziggurat the Sumerian and
Assyrian patterns for a temple tower are
combined. From the two stories reached
by a huge outdoor staircase, a spiral ramp
led up to four further floors. The ziggurat
of Babylon was the only tower in Mesopotamia in which the height was equal to the
length of the sides (300 feet). Xerxes
reduced the tower to ruins, and Alexander
the Great began to demolish it in order to
make room for a new structure—a project
that was cut short by his death.

171

The inner shrine of Ise on the Japanese
main island of Honshu. The temple pre-
cincts of Ise are the most revered of all the
sanctuaries of Shintoism. Only a century
ago Buddhist nuns and priests were not
allowed to enter it, and it is still without
the Buddhist statues, incarnations of
Shinto gods, that are found everywhere
else. The holy of holies is trebly guarded

by fence and palisade; it comprises treasure chambers for cultic objects and liturgical robes and the main hall in which lives the sun goddess Amaterasu, ancestress of the imperial house, who is symbolically represented by the sun mirror, holiest of the imperial insignia. The gateway to the inner shrine is opened only to high priests and the imperial family and its messengers. Tradition, however, has nothing against inspection from above. The Amaterasu shrine dates back to the beginning of the Christian era; since the seventh century it has been completely renewed, including the secondary buildings, every twenty years, if the times permitted. The sixtieth renewal took place in 1973, an exact copy being made of the old building and thus of all the previous buildings. Wood and straw are used exclusively, without either nails or mortar. The old building is divided into its component parts and sold down to the last chip on the devotional market; its wood is even reduced to toothpicks. In this way it pays for the new building.

172

A stupa in Sri Lanka (Ceylon). The stupa (the Singhaiese usually call it a "dagoba"), which is now the essential form of Buddhist religious architecture, developed from the pre-Buddhist princely tomb of ancient India. The Enlightened One himself instructed his disciples to honor kings and wise men with stupas. After his death, eight stupas were erected over holy relics of the Buddha, but King Asoka alone (third century BC), a pious champion of Buddhism, multiplied their number to tens of thousands. Built mostly over a holy relic, the stupa is always architectural sculpture without an interior. Above the stepped base, placed in an enclosure with four gateways, rises a hemisphere

which may also be bell-shaped or bulbous; the cubic superstructure at the top (sometimes containing the reliquary) is crowned by a cylinder and cone. This feature represents an umbrella—an ancient Oriental symbol of a ruler. To the believer the stupa symbolizes Parinirvana, the final, perfect Nirvana, the state of absolute salvation.

173

The radar radiotelescope of Arecibo, Puerto Rico. Its reflector, the largest on earth with a diameter of some 985 feet, is a stationary installation set in a depression in the terrain and directed toward the zenith. In contrast, the antennae on the triangular platform 525 feet above the ground of the punchbowl are swivel-mounted. Thus the apparatus can "see" 40 percent of the visible vault of the skies. (At the time this photograph was taken, the reflector consisted of fine-meshed wire netting; since then it has received an aluminium coating to improve its geometric definition). When Columbus set out to prove that the earth was round, he discovered Puerto Rico. Today astronomers penetrate to the limits of the universe from the tropical Antilles. The radio observatory of Arecibo records the murmurs of the quasars and pulsars. Nearer home, *El Radar* explores the ionosphere and the planets, whose orbits, thanks to Arecibo's tropical situation, are almost always in the field of vision of the telescope. In Arecibo it was shown that Venus, unlike the other planets, turns clockwise and not anti-clockwise. And that Mercury rotates on its axis in a mean period of fifty-five days—and not as previously believed, in eighty-eight days. Carl Sagan also uses the instrument to search for radio messages from extraterrestrial intelligences.

174

The Teatro Amazonas in Manáus, Brazil. Manáus on the Rio Negro, whose purple waters mingle with the brown floods of the Amazon nine miles downstream from here, did all in its power around the turn of the century to forget that it was surrounded by primeval forest. London provided the gentlemen's tailors, Paris the jewelers; cocottes and carriages were Continental. One of the trading kings in the jungle town even had his laundry sent to Europe by ship to be washed and ironed. Rubber was a monopoly of the Amazon, its prices were dictated by the dealers in Manáus—and world demand was rising. The spectacular theater, a copy of the Opéra in Paris, was the crowning symbol of the *dolce vita*. Caruso sang in it, Sarah Bernhardt acted there, and Pavlova danced. The big money also attracted lesser talents 800 miles up the Amazon, at that time an adventurous, fever-menaced journey. But even before World War I the rubber plantations of Southeast Asia had dethroned the collected rubber of the Amazon. Manáus went on for a while as if nothing had happened —but before long the tropical air with its smell of decay took over again where the fragrance of French perfumes had once prevailed.

175

The collegiate church of the Benedictine abbey of Maria-Einsiedeln, Switzerland, one of Europe's finest Baroque monastery complexes. The first monastery was built here by the Benedictines in the tenth century. The monastery and the pilgrim-frequented church were built in their present form in the eighteenth century— the sixth successive structures on this site. The fountain of St. Mary in the middle of the square is answered by a statue of the Madonna on the front gable.

176

Earth Art cut into the American earth: "Double Negative" by Michael Heizer on the edge of Mormon Mesa in Nevada. The cut is 30 feet wide, 50 feet deep and 1,640 feet long—including the middle part, which belongs to the concept. The tongue-shaped rubble dumps are also an integral part of the work. "The greatest sculpture in the history of Western art"— and its turning-point too—a prophet of Earth Art predicted in a reputable art journal. Nothing less than a record will do —and only fifty-five miles from Las Vegas; the West is in any case a courtesan who is ready to sleep with every art historian. "Double Negative" is undoubtedly the most labor-intensive work of Earth Art to date. Heizer, born in 1944, drilled, dynamited and bulldozed it in 1969/70 in a mountain he had bought a piece of. He used two tons of dynamite and moved over 200,000 tons of rock. The earth artist rejects the compulsions of the art trade— escapes from studios, galleries and museums and liberates his work from the pull of the consumer society. Heizer says: "One of the implications of Earth Art might be to remove completely the commodity status of a work of art and allow a return to the idea of art as . . . more of a religion."

177

"Spiral Jetty" by Robert Smithson, another American, in Utah's Great Salt Lake. Smithson, born in 1938, heaped up his spiral jetty in 1970 on a piece of land he had leased for twenty years. He kept account of exactly how much material he tipped into the lake to make the jetty, which is 490 feet long and 515 feet wide: 6,000 tons of basalt blocks and riparian rubble, 292 truckloads in all. This mysterious sign in the desolate landscape at the northern end of the lake reacts like an organic entity to the changing seasons. The photograph shows the spiral at high water. When the lake falls as a result of summer evaporation losses, the spiral is left dry, and the black blocks become encrusted with white salt crystals. At the same time the red coloring of the water is intensified, as the precipitation of the salt leads to an explosive multiplication of red algae. It should be added that the photograph does not do complete justice to the intentions of the artist. The drawn design was secondary to the experience Smithson had when walking along the spiral from the outside, of the swirl and suction of life, which spirals in irreversibly and ever faster toward the point of final truth—death. In 1973, when Smithson was searching from the air above Texas for a site for a new earth composition, a spiral ramp, he crashed to his death.

178
Cemetery of well-to-do Chinese near Medan, Sumatra, Indonesia. The striking and studied irregularity in the direction of the graves is the work of the geomancer, a wise man skilled in *fêng-shui* (Plate 118). In the upper right-hand corner a grave is just being dug. The bigger tombs are family vaults where wife and concubines lie beside the owner. In front of the horseshoe-shaped tumulus covering the coffin there is a space for funeral rites. Tombs of this kind extend over whole mountainsides in China. They reproduce, on a reduced scale, the ancient monumental tombs of the imperial house and of prominent persons. The mound symbolizes the mountain on which the soul of the deceased finally quits the earth to soar aloft to heaven. Thus the questioning never ends; the survivors go on asking in the name of the dead.

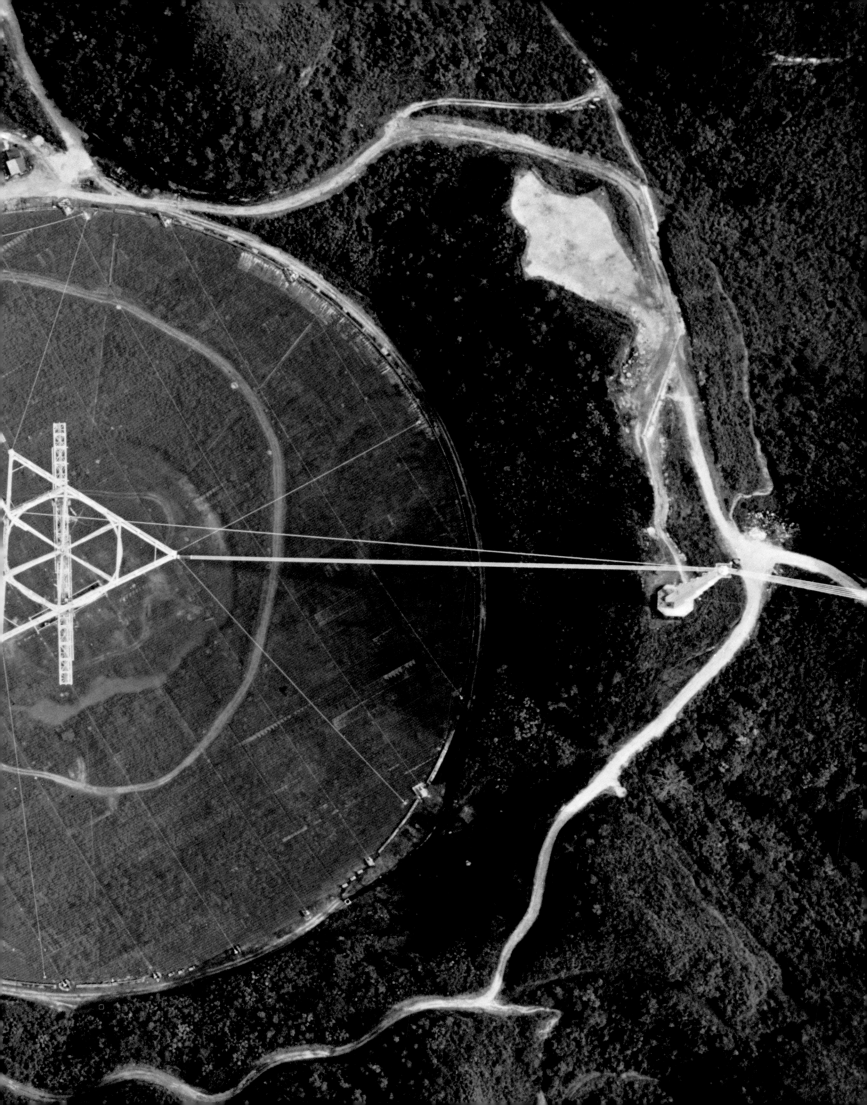

174

175

Pictures for the gods?

Of the many puzzles posed by prehistory probably the most curious is the fact that people without any means of leaving the ground created works, often of gigantic dimensions, which can only be seen or at least rightly appreciated from above. Such ground drawings and hill pictures are to be found in the Old and the New World, and especially in the Americas: in the chalky soil of the South Downs in England, in Peru's pampas and in Chile's Atacama Desert, on the terraces of the Colorado River and its tributaries, on the lava fields of the Sonora Desert in Mexico and in the grasslands and forests of the eastern half of North America.

Although the number of these works known today is surprisingly large, it is suspected that many more await discovery. We shall review briefly the most important known examples.

The desert-like plateaus in the catchment areas of the Rio Grande de Nazca y Palpa in southern Peru have ground drawings the profusion of which is unsurpassed in Latin America: a labyrinth of zigzag lines, triangles, rectangles, trapezoids, spirals, straight and serpentine lines, accompanied by figurative designs of birds, reptiles, dog-like quadrupeds, killer whales or fish, capuchin-like monkeys, an arachnid, and even some human faces. The drawing technique of the pre-Columbian artists was as simple as it was effective: by removing stones covered with the dark desert patina, they exposed the lighter ground. The dimensions of these designs can hardly fail to impress: a bird has a beak almost a thousand feet long, and some of the wedge-shaped triangles extend over miles. An attribution of the drawings to one or several periods of the Nazca culture (about 300 BC to 900 AD) has not yet been finally proven, but since the figurative motifs resemble designs on the beautiful Nazca ceramics it is reasonable to assume that they belong to the same culture.

The earth markings in the lower reaches of the Colorado—in today's states of California and Arizona and in the Mexican state of Sonora—show features analogous to those in southern Peru. Here too both figures and linear designs or "alignments" can be identified and are obviously interconnected. The Colorado tradition, however, spans 2,500 years. The earliest lines of stones must have been set up at the time of the early stone industries, at the beginning of the first millennium before Christ. The majority of the zoomorphic designs, however, belong to the end of the pre-Columbian era. If one assumes—as some archaeologists do—that the four-legged creatures depicted near Blythe, California, are horses, an outside date for these late drawings is automatically given: in North America the native horse became extinct ten thousand years ago, and the European horse was introduced in the west by the Spaniards around 1540. Many sites in the Colorado area are a type of palimpsest, for designs were made on the same ground for thousands of years. In the course of time, though, the techniques changed. The earliest pictures stand out in relief: stones on the desert surface mark their contours. In American archaeology these are often called "gravel sculptures." Later the artists adopted an intaglio technique: they sank the patterns in the ground by raking away the rock waste covering the surface. Often they gave the design additional relief by heaping up the debris along its edges. No agreement has yet been reached as to Indian cultures which produced these pictures. Their stylistic awkwardness is in striking contrast to the elegant designs of the Nazca.

A parallel to the gravel sculptures of the Colorado area can be found in the north of the continent, reaching far into Canada. Stones outline tortoises and snakes, bisons and human figures. Such "boulder mosaics"—an unhappy designation for pure outline drawings—are the pride of Whiteshell Provincial Park, 120 miles northeast of Winnipeg, in the Canadian province of Manitoba. They are probably the work of Chippewa Indians, but the exact time of their creation is uncertain. Estimates oscillate between centuries and millennia. An attempt by scientists from the University of Manitoba to date these designs by studying lichens on the boulders has so far been unsuccessful. (Comparable boulder designs are still being made today by Australia's aborigines on rounded mountaintops, particularly in the north of the continent. The ground enclosed by single stones and stone heaps is reserved for tribal rites. The shapes of some of these ceremonial areas also suggest animals when seen from above, though only vaguely drawn.)

The artificial mounds of various American Indian cultures are concentrated in the region east of the continental watershed, and especially numerous over the huge area from the Rocky Mountains to the Appalachians and from the Great Lakes to the Gulf of Mexico. These giant mounds, mostly made of soil, served in some cases as burial tumuli and in others as platforms for ritual worship. Some of the mounds are in the shape of snakes, tortoises, birds, bears, lizards, foxes or bisons—imagination, of course, plays a large role in their identification. In contrast to the earth pictures, two birds (eagles?) found in today's Georgia were formed by the heaping-up of white quartz stones. The tribes responsible for these mounds did not always use them for burials, and even where they did the objects interred with the dead are now mostly missing, though abundant in the tumuli proper. This circumstance makes a reliable determination of their age difficult. The majority of these mounds were probably made after 500 AD, but the Great Serpent in Ohio, the most spectacular of all, is attributed by archaeologists to a culture existing there between 1000 BC and 400 AD. In the nineteenth century the mounds in the valleys of the Mississippi, the Missouri and the Ohio were a distinguishing feature of the landscape. Grave plunderers and well-meaning amateur archaeologists, together with the plow, have since largely destroyed this legacy of the past.

Despite this, it is estimated that approximately 5,000 mounds have survived, almost all in Wisconsin. The only big group outside Wisconsin is the Effigy Mounds National Monument (in Iowa, across the Mississippi, opposite the small village of Prairie du Chien, Wisconsin) and is under federal protection. The settlers of the nineteenth century, with their characteristically low opinion of "Redskins," attributed these astonishing earthworks to a non-Indian people. Excavations since then have, of course, removed any possible doubt as to their Indian origin.

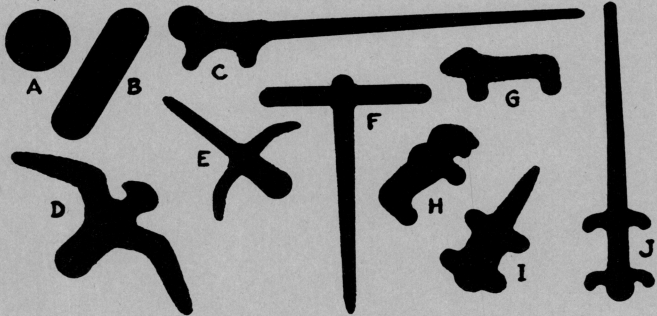

Mounds in the valleys of the Mississippi and Ohio. A and B are nonfigurative, the rest zoomorphic: panther, eagle, aquatic bird, bird, bear, buffalo, tortoise, lizard— with room left for the imagination!

A brief recapitulation must suffice for the earth drawings found in Britain. In the hilly chalkland of southern England, where turf covers the chalk like a skin, the artist worked by cutting out strips of sward. Students' jokes and gentlemen's folly, the boredom of garrisons and the gratitude of city fathers for marks of royal good-will have enriched the Downs since the eighteenth century with a dozen such works of grass art. The impulse, however, was provided by three works whose origins lie in the dawn of English history: the White Horse of Uffington and the two giants. The White Horse on the edge of the Berkshire Downs must no doubt be seen in relation with the so-called Uffington Castle, a fortress built in the Iron Age on the hilltop immediately above the horse. Similar stylized representations of horses on Celtic coins of the Iron Age would suggest an age of roughly 2,000 years for this work. The English archaeologist Stuart Piggot, an authority on the subject, thinks that the Giant of Cerne Abbas in Dorset is connected with the Hercules cult which spread in Britain under Emperor Commodus in the second century AD. But it is possible that the figure received its knotty Herculean club at a later date. The hill with the giant certainly lies in a region full of relics of the New Stone Age, the Bronze and the Iron Ages. The second giant, the Long Man of Wilmington in Sussex, has no less claim to a prehistoric past if the monuments strewn around him are anything to go by. His age is extremely difficult to determine, however, because of a nineteenth-century "restoration" which, it is presumed, altered the original shape both of the giant and of the objects he is holding.

If the dating of these unusual prehistoric pictures raises many problems, the question of *how* they came into being is far more difficult still.

The American Paul Kosok, the discoverer of the Nazca figures, called the area — thirty miles in length from north to south — "the biggest astronomy book in the world." The linear patterns might be interpreted as sight lines for a gigantic astronomic almanac of the sun, moon and stars. Maria Reiche, the greatest living authority on these pictures, has studied them for decades and subscribes to

283

this theory: the (superimposed) drawings record the steadily growing astronomic and calendrical knowledge of the priest-astronomers of those times — who are admittedly hypothetical. Lines that obviously do not refer to the solstice or equinox may have served to indicate planets or stars on special days in the calendar. Several years ago the Stonehenge researcher Gerald S. Hawkins carried out a computer analysis of the Nazca lines. His findings disappointed all who had read into them a piece of archaeo-astronomic lore: the computer was unable to discover any kind of astronomical meaning in the majority of the lines, and the number of calendrically significant lines did not exceed the quota allowed by statistical chance. The line systems and animal figures on the Peruvian pampas had once more become a mystery. (Maria Reiche's criticism and doubts with respect to Hawkins's method are expressed in the catalogue of the exhibition of Peruvian Earth Signs staged in Munich's Kuntsraum in 1974.) American archaeologists have sought a key to the meaning of the giants in the Colorado region in a recent act of the Pima Indians: the latter scratched the shape of a giant into a mesa near Sacaton, Arizona, to commemorate the destruction of a child-eating monster. In the myths of several tribes in the Southwest there are both well-meaning and evil giants with whom the Indians themselves connect these ground pictures. Other archaeologists point to the similarity of the earth drawings to the rock carvings of the American Southwest and speculate on their magic function for the hunter. Indians now living, familiar with the traditional lore of their tribes, make fun of the profound interpretations the white man now reads into their rock designs. As for the mounds in the East of the U.S.A., a form of totemism seems probable. Where the mounds contain tombs, the dead were always buried at points corresponding to vital parts of the depicted animal, particularly in the region of the head, the heart and the groin. Even the White Horse of Uffington can be interpreted as

284

some kind of totem or at least as a tribe's emblem. A native fertility cult existing long before the Roman invasion of Britain may have been responsible for the creation of the Giant of Cerne Abbas.

All these considerations, however, miss the cardinal problem: why were the pictures executed in a size that made them practically invisible to the earthbound observer? We can give no answer beyond noting that Man, imitating the psychic pattern of the parent-child relationship, has always verticalized his religious world, looking up with hope and down with fear. "Explanations" interpreting, for instance, the overlapping triangles and trapezoids of the Nazca as a prehistoric space airport where divine culture-bringing astronauts landed need not be taken seriously. The best-sellers they have begotten are signs of the times, of more interest to psychiatrists and sociologists than to prehistorians.

Two assumptions, however, may be made about the earth drawings and picture mounds of the Old and New Worlds. They are in the first place the result of a group effort. Secondly, they served some kind of cultic purpose. The soil displacement necessary to create the mounds would be unthinkable without the efforts of large numbers of people. The builders of these mounds obviously lived in small scattered groups for most of the year and gathered together only during the sowing and harvesting seasons. Only at these meetings was the labor pool sufficient for an enterprise as ambitious as the building of these mounds. Furthermore, translating a sketch into the correctly scaled proportions of the final giant designs on the pampas of Nazca — calibrated ropes were probably used — would have been impossible without an army of helpers. The same is true of the drawing of the lines, the straightness of which still wrings admiration from land surveyors. Perhaps the Nazca figures were paced out by processions. The contours consist of a single endless line and thus made it possible for marchers to return to their starting point without turning back.

But it is not only the purpose of these earth pictures and mounds that is to be sought in a cult—their creation was ceremonialized as well. Perhaps it is not altogether accidental that modern landscape artists, the exponents of Land Art, who draw their compositions in the mesas of Nevada, the cloverfields of Wisconsin and the Great Salt Lake, the dunes of the Sahara and the slag fields of Mexican volcanoes, are concerned less with the permanence of their work than with the actual act of creation. Yesterday's cult and today's happening! In Peru, Mexico, California and Arizona the desert climate has preserved these pictures. The Atlantic climate in the south of England is less hospitable to the grass compositions: rain washes out the exposed chalk and the wounds heal fast in the lush grass carpet. Despite this, the giants and the White Horse of Uffington have survived for thousands of years. The local populace "groomed" the Uffington Horse every seven years: the weeding-out of undesired grass and the replacement of washed-out chalk, again to the rhythm of a holy number, have saved the figure till our own day. This cleaning ritual, to judge from a description that has come down to us, used to be accompanied by dances and games, thus re-enacting the original rites in the gayest of spirits. We do not know how many giants and horses have vanished elsewhere for lack of such loving care.

179

The Long Man of Wilmington: grass figure in Sussex, England. The giant is 230 feet high, and the two rods he is holding in his hands slightly overtop his figure. The ditches which form his outline in the grass are over two feet wide. For centuries the Long Man has been the subject of speculation: he has been variously interpreted as Apollo, Hermes, Baldur, Woden, Beowulf, Thor, the Hindu goddess Varuna, Mohammed or the Apostle Paul. But it is possible that the figure merely served to commend the Benedictine priory of Wilmington to passing pilgrims . . . . Of all ancient earth carvings, the Wilmington giant can be seen best from ground level because of the steep gradient of the slope. Unfortunately, those who restored the figure in 1874 took some liberties, and doubts about the giant's exact appearance prior to that date now make a reliable identification impossible.

180

The Cerne Abbas Giant: grass figure in Dorset, England. From club to foot the giant measures 180 feet. The line incisions in the turf have a triangular cross-section; they are approximately twenty-four inches deep and wide. For the creation of this figure almost twenty tons of turf and chalk had to be removed. Whether he is seen as a Romano-British Hercules of the third century AD or a far older fertility god *a la* Priapus, the giant remains the most astonishing work of grass art in the south of England. Outbreaks of prudery, both mild and acute, have not, of course, spared the ithypallic figure. An issue of *Gentlemen's Magazine* in 1764 published the first picture of the giant. To be on the safe side, the journal branded him as a work of the Papists and entered in its reproduction all bodily measurements save one. Ten years later, the historian J. Hutchins, in his history of Dorset, completely omitted the

offending member. Night-excursions have been organized for the express purpose of contriving a figleaf made of jute and paper to cover the nakedness of the shocking fertility idol. But every such attempt at obfuscation has been answered by alert local citizens with a counter-expedition to restore the Priapic insignia. Childless women of the neighbourhood have secretly visited the giant in hopes of redress. The era of pill, pop and porn, however, has replaced ancient fertility beliefs by a snapshot of the girlfriend standing on the giant's genitals.

181/182

Pictures engraved by native Americans in old river terraces of the Colorado near Blythe, California. In 1854 the geologist W. P. Blake, who was exploring the route for a railway connection from the Mississippi to the Pacific, first noticed a ground drawing in the Colorado area. But the figures were only really discovered (in the sense that they attracted scientific attention) in 1932, when a pilot sighted a number of giant figures near Blythe. An exact survey of the region revealed three groups of designs incised in the earth. The first (Plate 181) comprises a giant, a quadruped and a spiral shape. The giant— or ogress, as her sex seems originally to have been clearly marked—measures ninety-one feet from head to foot; the span of her arms exceeds sixty-nine feet. The quadruped (a horse?) is forty-three feet long and thirty feet high to the shoulder. The giant and the animal are of the same age, but they have been superimposed on older drawings. An ancient circle, which may have been inscribed in the desert soil by dancing feet, is today visible only where it intersects the legs of the giant (just below knee height). The spiral shape near the animal also belongs to an earlier drawing era. The group is relatively easily accessible from a nearby road and has therefore been fenced off. The tracks of motorized visitors have enclosed the figures in a modern ring (the dance is now St. Vitus's). The second group (Plate 182), hardly half a mile away from the first, again consists of a giant human figure and a four-legged animal. The giant (female once more, as was quite clear before vandals began to disfigure her) is 168 feet high, with an arm span of 157 feet—the largest of the Blythe figures. In 1952 her long hair was still visible. The accompanying animal is also larger than the other quadruped. It is fifty-two feet long and forty-two feet high. (Aside from these two main groups, completely on his own, is a third giant, 105 feet high; in this case stones have been used to represent male genital organs.)

183
Not far away, the giant near Sacaton, Arizona (in the Gila River reservation of the Pima Indians). The giant is 175 feet long; his body consists of a single furrow, 20 inches wide; heaps of stones mark his extremities and important parts of his anatomy. Next to the giant appears a smaller figure (a child?) designed in the same style. The Pima Indians call the giant Hâ-âk-Vâ-âk, which means "Hâ-âk lying." In the legends of the Pimas, Hâ-âk is a child-eating monster (of female gender) which was killed by an Indian Theseus named Elder Brother at the request of the frightened population. In memory of their liberation from this monster the Pimas made this shrine, which is still of religious significance in tribal life. The memorial was described for the first time around the turn of the century and is probably not much older. But many American archaeologists regard it as the key to the meaning of the other giants near Blythe.

184
America's Great Serpent: Serpent Mound State Memorial in Adams County, Ohio. The snake that coils over a ridge beside Brush Creek (frozen over at the time the photograph was taken) is over 1,300 feet long. The reptile is portrayed with open mouth, in the process of swallowing an egg(?). The outline of the snake was first marked with small stones and lumps of clay. Large quantities of yellow clay were then obtained from a pit nearby in order to mold the snake's body over the markings. The work contains no reference of any kind to its age or the culture of its creators. A burial mound not far from the Serpent is attributed to the Adena culture; in view of this contiguity the Serpent Mound is now credited to the same race (known to have lived in Ohio until 400 AD). The Great Serpent established a precedent in the history of American archaeology as one of the first conservation projects. It was endangered shortly before the Civil War when a tornado mowed down the forest that had so far hidden it. Farmers cut down the last remaining trees and set about plowing the land. A Harvard archaeologist sounded the alarm, and "several of Boston's noble and serious ladies sent out a private circular. . . ." The contributions that poured in sufficed for the purchase and donation of the Memorial.

185
Rock Eagle Effigy Mound near Eatonton, Georgia, U.S.A. The bird—an eagle in one interpretation, a buzzard in another—is made of white quartz stones which had to be brought from afar by its Indian creators. It measures 119 feet from wing tip to wing tip. An observation tower has been erected at its foot to enable present-day visitors to see the bird properly. A second bird mound is located in the same area.

186

The so-called Little Bear Hill in America's
Effigy Mounds National Monument, Iowa.
The outlines of the Little Bear have been
artificially retraced.

187

Horse and Rider near the village of Osming-
ton in Dorset, England. The gigantic war-
horse with rider is 328 feet high. Thomas
Hardy (in "The Trumpet Major") connects
it with the battle of Trafalgar. Others regard
it as the leisure production of a garrison
waiting in vain for an invasion by Napo-
leon's troops. Most probably, however, the
picture honors George III, for whose
patronage the nearby town of Weymouth
had reason to be grateful. The work
originated around 1815. No other White
Horse in the South of England carries a
rider.

188

The White Horse of Alton Barnes in Wiltshire, England, which dates from around
1812. Like all the White Horses of Wiltshire, it trots to the left. It is roughly as
high as it is long, approximately 164 feet.
The aerial photograph unduly accentuates
its height; when the work is looked at from
ground level, the inclination of the slope
serves to correct the proportions. Like the
other Wessex horses, that of Alton Barnes
is a descendant of the Uffington Horse
(next picture), which first suggested the
idea. Stylistically, however, the later creations show the influence of popular horse
painters of the day. It was part fun, part
fad, that resuscitated prehistoric grass art
in England in the eighteenth and nineteenth centuries. Country gentlemen, cler-
gymen, doctors, shoemakers and firemen
all joined in the game. A gentleman farmer
by the name of Robert Pile commissioned
a painter's journeyman to design the horse
of Alton Barnes, but the fellow disappeared
with the cash advance. Thus Mr. Pile had

to fork out a second time, though he later
had the satisfaction of seeing "Jack the
Painter" hanged.

189

The White Horse of Uffington in Berkshire,
England. Overall length: 360 feet. Its adventurous stylization—two legs in search
of a barrel—has aroused a suspicion in
some present-day observers that it is
meant to be a fox—or a dragon, or a grey-
hound, or even an ichthyosaurus. In contrast, such doubts were never cast on the
horse in the old sources, and in a four-
teenth-century treatise on the wonders of
Britain it ranks immediately after Stonehenge (Plate 168). A fence now protects
the horse from souvenir and fortune
hunters (anyone who turns around blind-
fold in the horse's eye three times is
supposed to be in for good luck). The
horses in the chalk—the archetype of
Uffington as well as the newer creations—
are landmarks visible at great distances
from the air. They are included on pilots'
maps as orientation aids. The British War
Ministry in World War II therefore had
every reason to have them obscured:
green paint, turfs and twigs camouflaged
them for German airmen from 1940
onward.

190

Pebble designs by Australian aborigines
(without any figurative content) near
Cannon Hill in Northern Territory. These
lines of single stones or heaped pebbles
can be found throughout Australia. The
aborigines use them to mark areas reserved for ritual ceremonies. The act of
demarcation, the lining-up of the stones,
itself has ritualistic significance.

191

Remains of a large "boulder mosaic" in
Whiteshell Provincial Park, Manitoba,
Canada. The boulders are arranged on a

289

granite outcrop in a clearing. Chippewa Indians probably assembled such figures at places of sacrifice. Alexander Mackenzie, who navigated Winnipeg River in 1793, observed similar customs among the Chippewas of that time.

192

Figure incised in the Pampas, between Nazca and Palpa in South Peru. Maria Reiche interprets the 165-foot drawing as an ideogram for a monkey—with skull, two ears, two hands and nine fingers. The more explicit drawing of a spider monkey near by—also with nine fingers—supports her theory.

193

Nazca earth figure: an arachnid measuring 151 feet in length. Gerald S. Hawkins believes that the striking prolongation of one of the walking legs (of the third pair) identifies the creature as a member of the order of Ricinulei, an inhabitant of the virgin forests of the Amazon. The males of this spider species use their third pair of legs as a sexual organ. On the strength of this, Hawkins speculates on the possible role of this picture in a fertility cult.

194

Nazca earth figure: a giant bird, measuring 443 feet from tail to beak. Maria Reiche identifies it as a flying frigate bird seen from below.

195

Nazca earth figure: a giant reptile, lizard or crocodile, almost 625 feet long, cut in two by the Pan-American Highway (lower left-hand corner of photograph).

196

Nazca earth figure: representation of a plant (?). Maria Reiche sees it as seaweed. Bottom left the Pan-American Highway once more, which is the curse of the Nazca earth engravings. Not only were many of the designs destroyed during the construction of the road, but too many users of it cannot resist the temptation to enter their signatures with tire-tracks in this book of pre-Columbian symbology—marks of contemporary stupidity amongst the ideograms of prehistoric wisdom.

197

Nazca earth figure: a flower, 131 feet long. Maria Reiche stresses the orientation of the stem toward the west. Could it be an equinoctial line?

198

Nazca earth figure: a drawing of a bird (a colibri?), 328 feet long. The beak of the bird ends at a line which—by chance?—points toward sunrise at the time of the winter solstice.

199

Nazca earth figures: straight lines and surfaces near the escarpment of the desert plateau. The larger triangle has a base of almost 395 feet, the smaller one of over 100 feet. The unbridled imagination has turned the nonfigurative motifs in the wealth of Nazca earth figures into landing runways and assisted take-offs for extra-terrestrial space pilots. It is sad to have to suppose that these super-intelligences from some distant planet had not made a little more progress in their aviatic technology. . . .

200

Nazca earth figure: representations of human forms on a slope. In contrast to the earth designs on the desert plateau, which always consist of a single un-interrupted line, these slope drawings are made up of separate strokes—perhaps because of the technical difficulty of engraving designs on the steeply sloping surface, perhaps because they are earlier or later productions.

179

180

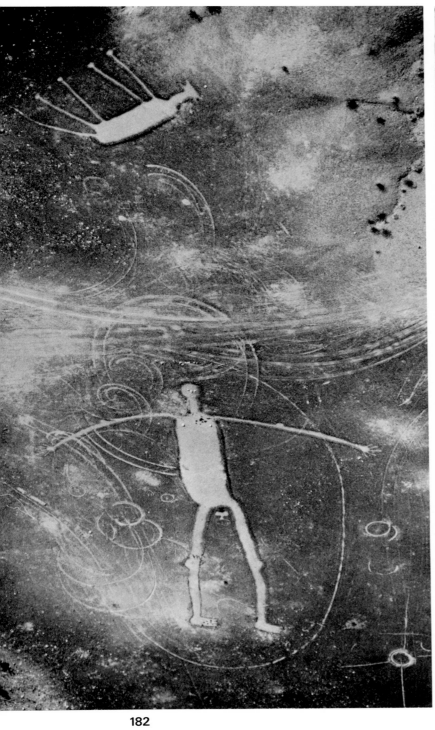

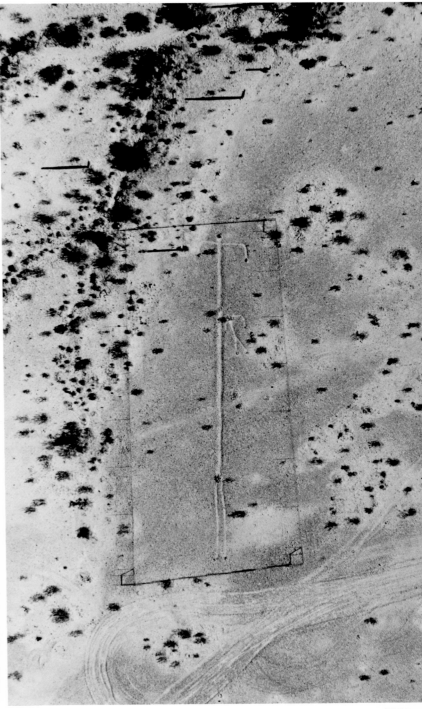

182   183   184 ▶

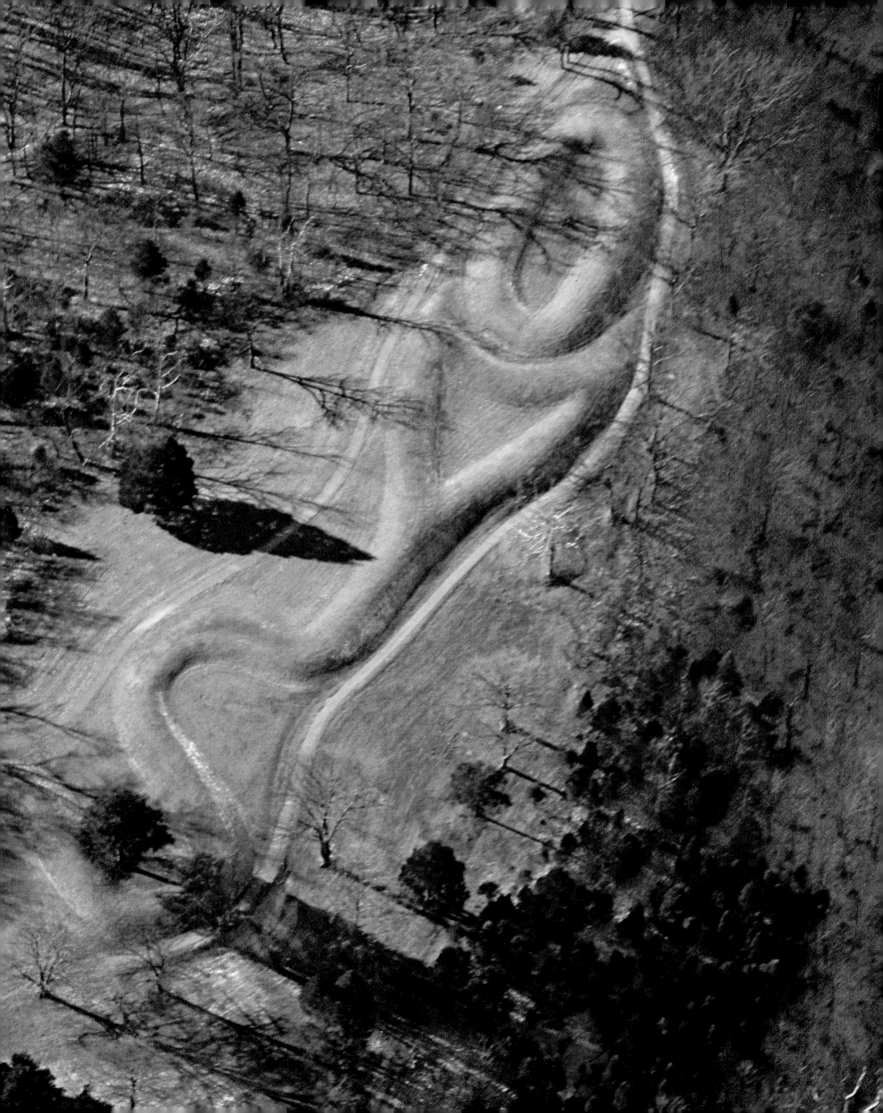

185-188

189

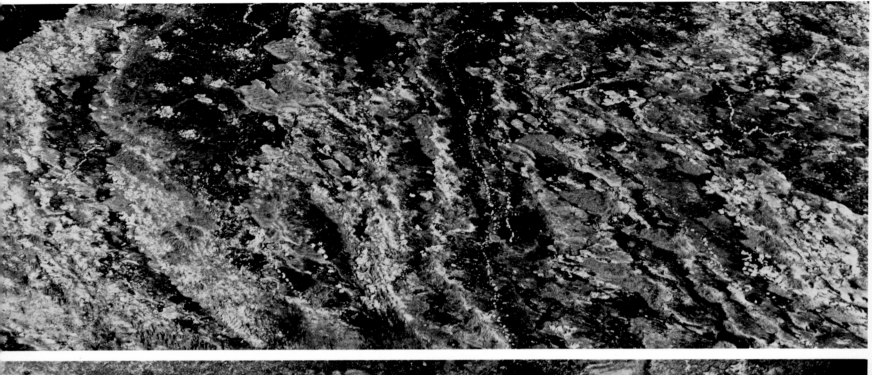

192

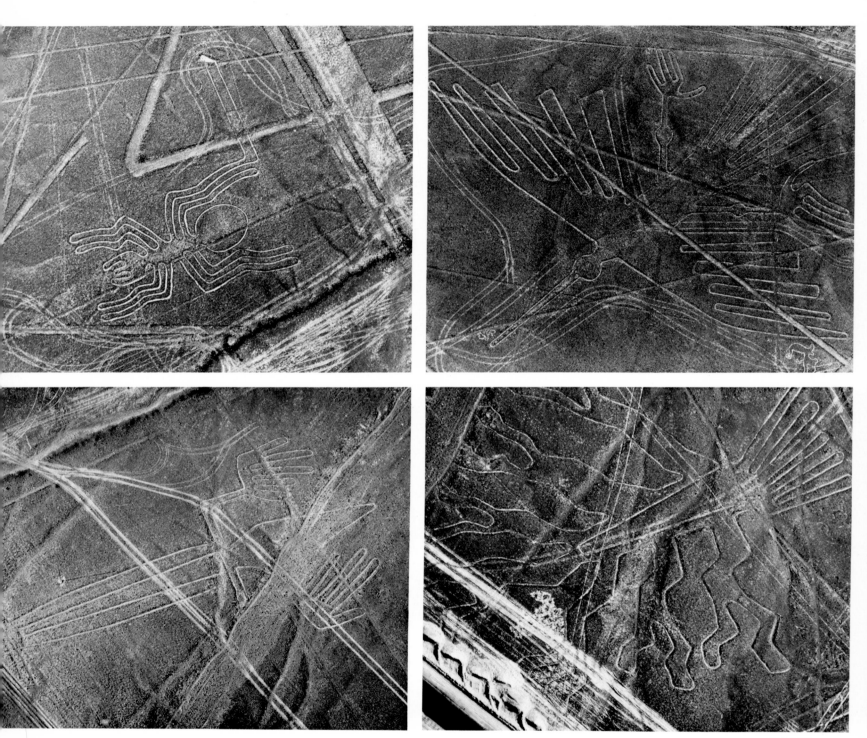

200

List of plates

306

Calligraphy of the industrial age.

Saltworks as things of beauty.

308

309

Pictures for the gods?

Notes

The Photographs

To collect the photographic material for this book, the author spent more than a thousand hours in small chartered planes flying over fifty-nine countries and territories. In the final selections of the illustrations, subject matter came first and geographical variety was only a secondary consideration. Even so, the examples included represent forty countries and territories on all seven continents.

Equipment

Nikon cameras equipped with motor drive and Nikkor lenses with focal lengths from 28mm to 180mm were used. The motorization of the camera made it possible to select subjects and angles in fractions of a second while approaching and flying over the target. This resulted in a considerable saving, as expensive helicopters were not needed.

Film Material

Ilford FP-4 was used for the black-and-white photographs, Kodachrome II or Kodachrome 25 for the color shots. The resolving power and sharpness of Kodachrome films, together with modern reproduction techniques, yield results in the miniature (35mm) photograph which make cameras of the next larger negative size unnecessary.

Publication Permits

The Austrian Federal Ministry of Defense released Plates 37, 49, 145 and 146 for publication with the note Z1.10.831-RabtB/74.

Illustrations in the Text

The Icelandic plan of Jerusalem in the chapter "An Archetypal Settlement: the Round Town" was drawn by Albert Gerster from a manuscript in the possession of Copenhagen's University Library (reproduced by W. Müller in his book *Die heilige Stadt*). For the plan of the types of mounds in the valleys of the Mississippi and the Ohio in the chapter "Pictures for the gods?" I am indebted to the administration of the Effigy Mounds National Monument. The sources of the remaining illustrations in the text are acknowledged in their respective captions.

Literature used

The titles listed below are meant as an expression of the author's gratitude to writers whose books and papers helped him in the compilation of the notes on the photographs—they are not a bibliography, not even a select one. Where the information given on the photographs is widely known or easily accessible, references to the literature have been omitted altogether. Persons who assisted the author by supplying information orally or by letter are listed, with thanks, at the beginning of the book.

FLIGHT OVER THE AFAR
Holwerda, J. C. and R. W. Hutchinson: "Potash-Bearing Evaporites in the Danakil Area, Ethiopia," in *Economic Geology*, Vol. 63, No. 2 (March/April, 1968).

A ROOF OVER ONE'S HEAD
Le Corbusier: *Aircraft* (*The New Vision*, Vol. 1.) London, 1935. Hofer, P., "Die Stadtgründungen des Mittelalters zwischen Genfersee und Rhein," in *Flugbild der Schweiz*, Berne, 1963. Westphal-Hellbush, S.: *Die Ma'dan*, Berlin, 1962.

AN ARCHETYPAL SETTLEMENT:
THE ROUND TOWN
Creswell, K. A.: *Early Muslim Architecture*,

Oxford, 1932–40. Moholy-Nagy, S.: *Die Stadt als Schicksal*, Munich, 1970. Muller, W.: *Die heilige Stadt*, Stuttgart 1961. Vogt, A. M. *Russische und französische Revolutions-Architektur*, Cologne, 1974.

CALLIGRAPHY OF THE INDUSTRIAL AGE
Cook, W. H. "Stabilizing Gold Mine Slimes Dams and Sand Dumps by means of a Vegetative cover." Lecture to the South African National Association for Clean Air, 1970. (Available through the Chamber of Mines Services, Johannesburg). Groves, J. E.: "Reclamation of Mining Degraded Land," in *South African Journal of Science*, Vol. 70 (October, 1974).

THE FARMER AS ARTIST
Coe, M. D.: "The Chinampas of Mexico," in *Scientific American*, July, 1964. Evenari M/Shanan, L/Tadmor, N., *The Negev*, Cambridge, Mass., 1971. Needham. J, *Science and Civilisation in China*, Vol. 2 of *History of Scientific Thought*, Cambridge, 1956. Ron, Z.: "Agricultural Tetraces in the Judean Mountains," in *Israel Exploration Journal*, Vol. 16 (1966). Schumacher, E., "Photo-chemische Speicherung der Sonnenenergie," in *Neue Zurcher Zeitung*, No. 114, May 21, 1975. Troll, C.: "Qanat-Bewässerung in der Alten und Neuen Welt" in *Mitt. d'Österr. Geograph. Ges.*, Vol. 105 (1963). Whyte, R. O.: "Evolution of Land Use in South-Western Asia," in *A History of Land Use in Arid Regions*, Unesco, Paris 1961.

ARCHAEOLOGISTS TAKE TO THE AIR
Crawford, O. G. S.: *Wessex from the Air*, Oxford, 1928; and: *Air Photography for Archaeologists*, Ordnance Survey Professional Papers, London, 1929. Dolenz, H.: "Luftbild und Archäologie," a letter to Johann Dominikus Prunner von Sonnenfeld, published in *Kosmos*, Stuttgart,

November, 1973. Newhall, [...] *Camera*, New York, 1969. [...] "Kuntswissenschaftliche Arbeit während des Weltkrieges in Mesopotamien, Ost-Anatolien, Persien und Afghanistan", in *Kunstschutz im Kriege*, Leipzig, 1919. Scollar, I.: *Archäologie aus der Luft*, Düsseldorf, 1965. Vorbeck, E. and L. Beckel: *Carnuntum*, Salzburg, 1973. Wiegand, Th., "Denkmalschutz und kunstwissen-schaftliche Arbeit während des Weltkrieges in Syrien, Palastina, und Westarabien," in *Kunstschutz im Kriege*, Leipzig 1919. Wise, J.: *Through the Air*, Philadelphia, 1873.

BIBLICAL SITES AND CITIES FROM THE AIR
Dever, W. G. (*inter alios*): "Further Excavations at Gezer, 1967–1971," in *The Biblical Archaeologist*, Vol. 34. Cambridge, Mass. (1971). Segal, A.: "Herodium," in *Israel Exploration Journal* Vol. 23. (1973).

MONUMENTAL QUESTION MARKS
Hawkins, G. S. *Stonehenge Decoded*, New York, 1965. Killer, P. "Leitfossilien der Zeitkunst 1945–1973," in *DU*, April 1973. Tomkins, C. *"Onward and Upward with the Arts,"* in: *The New Yorker*, February 5, 1972.

PICTURES FOR THE GODS?
Folsom, F.: *America's Ancient Treasures*, New York, 1971. Hawkins, G. S.: *Beyond Stonehenge*, New York, 1973. Marples, M.: *White Horses and Other Hill Figures* London, 1949. Reiche, M.: *Geheimnis der Wüste*, Stuttgart-Vaihingen, 1968. Rogers, M. J.: *Early Lithic Industries of the Lower Basin of the Colorado*, San Diego, 1939; and: *Ancient Hunters of the Far West*, San Diego, 1966. Steinbring, J., "The Tie Creek Boulder Site of Southeastern Manitoba," in *Ten Thousand Years, Archaeology in Manitoba*, Winnipeg, 1970.

312

4

73

64
142
168
179
180
187

4
16

99
100
125

86

42

191

74
177

129-141

186

6

78

21
95

2

176

184

79
90
106

5

181
182

3

61
183

63

82

43
55
70
72

83
92
93
94
96

20
117

185

124

16

60

91
115

173

84

98
46

53

62

45

174

192-200

57

68

65
120

123

152

113
114
119

Südpol
Pôle du sud
9
10
South Pole